The Beekeepers
Annual 2015

THE BEEKEEPERS ANNUAL
IS PUBLISHED BY
NORTHERN BEE BOOKS
MYTHOLMROYD,
WEST YORKSHIRE

PRINTED BY
LIGHTNING SOURCE, UK
ISBN 978-1-908904-63-8

MMXIV

EDITOR, JOHN PHIPPS
NEOCHORI, 24024 AGIOS NIKOLAOS,
MESSINIAS, GREECE
EMAIL manifest@runbox.com

SET IN HELVETICA LT BY D&P Design and Print

Cover: Honeybees waste nothing. All their products are valuable and recycled when possible.
Here a bee has discovered a sheet of foundation in an empty hive and
is collecting it for its own colony. *(Photo: John Phipps)*

The Beekeepers
Annual 2015

CONTENTS

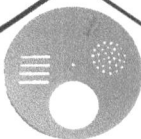

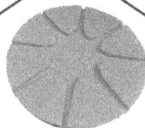

FOREWORD

John Phipps

September 2014

August 2014, the beginning of the First World War, almost exactly a century ago, has been commemorated in many ways. Looking back, it is shameful to see how the lives of millions of people were destroyed or horrendously changed by a war which could have been avoided. The over-used epithet describing the event as "The War to End all Wars" as if to vindicate the loss of so many lives sadly turned out to be untrue as both subsequent and recent events have shown.

In this edition of The Beekeepers Annual, we have focussed, in the main, on beekeeping at the end of the 19th Century and the beginning of the 20th Century with extracts from books from that time, the contents of which are still of relevance today. The plates for each month in the Calendar section are advertisements, similarly taken from old beekeeping books, as they give us some idea of who the beekeeping suppliers of that era were and the types of hives and other equipment they sold, including prices. Sadly, of the range of manufacturers and equipment suppliers of that time none still remain.

More pertinent to the war itself, Stuart Ching, Editor of the Notts BKA Newsletter, has provided us with a lengthy piece on the problems with the supply of sugar and what this meant in real terms for the householder and beekeeper.

Important lessons were undoubtedly learned during the war, lessons which we should pay heed to today. Most importantly, people realised the value of all commodities and any form of waste was seen as helping the enemy - for example, wasting food meant that fewer supplies could be send to the troops on the front line, or fed to starving refugees, i.e. people displaced from the theatres of war.

I mention this particularly, as it is estimated that about 50% of the food now produced in the world is wasted. This means, too, that all the energy put into its production, transportation and storage is similarly wasted - using unsustainable sources of energy. Agriculturists and ministers responsible for food do little to address this problem and seek instead to increase production for the rising population by turning to new technology - like the cultivation of genetically modified crops, which are mostly unwanted by consumers. It is important to conserve what is already produced and stop the waste. Maybe only events like the war we are now commemorating will bring this message home.

John Phipps
September 2014

1914 AND ALL THAT...

John Phipps

During the First World War about 2 500 British merchant vessels were sunk or damaged beyond repair.

As an island nation and having a good navy, Great Britain for over a thousand years has been able to protect its shores from its enemies. Aerial warfare beginning in the First World War, however, made Britain more vulnerable to enemy attack, but the greatest threat to the country was the procurement of much needed food and other valuable resources as merchant shipping was constantly targeted by German U-boats and other vessels of war.

Most of the naval casualties were caused by German U-boats.

Between August 1914 and November 1918 an estimated number of 2,483 ships were lost at sea, though this figure does not include the considerable number of trawlers which were sunk.

The totals for each year:
1914 -   64
1915 -   278
1916 -   395
1917 - 1204
1918 -   542

Records for individual ships show that they were torpedoed without warning; or captured by U-boats and sunk by torpedoes, bombs, gunfire or scuttled. In some cases the captured ships would be taken into one of the enemy's naval bases as a trophy, with all the cargoes then being plundered. A particular menace to shipping was the "Mowe". Disguised as a merchant ship, this raider was responsible for the sinking of forty ships, either directly, or by the mines which she lay in many of the shipping lanes.

Although the figures given above show an increase in the number of ships destroyed, the total would have been considerably higher if they hadn't sailed in large convoys or had support from the air which was able to locate and destroy enemy submarines.

A very interesting and unusual scheme to try and deceive enemy ships was conceived by the artist Norman Wilkinson. Instead of the ships being painted in such a way as to camouflage them, he suggested that the exact opposite should be tried instead. Thus some ships were painted in bold contrasting colours with large cubist type patterns which were meant to confuse the enemy. Seen from the distance the patterns on the bows of the ship and the funnels would make the ship look as if it was going in the opposite direction.

Amongst the many staples badly needed was sugar (see the article by Stuart Ching), for most of Britain's supply was sourced from sugar cane grown in its colonies, or the more inferior sugar from sugar beet grown in western Europe, including Germany.

6

Certainly, the lack of sugar was one of the problems faced by beekeepers at this time, but there were other things, too, which affected beekeeping as a commercial enterprise or rural household activity. Many beekeepers were called upon to fight, not all of them sadly to return, with the colonies succumbing soon after unless they were taken over by someone in the family or community; the "Isle of Wight Disease" was still taking its toll; and basic equipment for beekeeping was in short supply. This also applied to catalogues and books, as savings had to made on paper, another commodity for which the country relied on imports.

Of course, all efforts were made to increase home production of food, and for the first time, many women were drafted into the land army; it is not known how many women actually took up beekeeping. Sadly Fraser's book "History of Beekeeping in Britain" first published in 1958, ends in 1900; another half or even a quarter of a century would have thrown light on these interesting and difficult years which covered two world wars. A few books were published in England which aimed to encourage the taking up of beekeeping. These included:

"Beekeeping simplified for the cottager and smallholder", William Herrod-Hempsall,1915 (just 48 pages)

"Bees and their management", William Herrod-Hempsall, - it being the first chapter (63 pages) of "Livestock of the Farm' Vol. 6, C Bryner Jones 1915 - 1916

"Productive bee-keeping: modern methods of production and marketing of honey," Frank C Pellet, 1916, (302 pages)

"Bee-plants and their honey," Mrs M Grieve, 1917, (17 pages)

"Bee-keeping in war-time," William Herrod-Hempsall, 1918, (32 pages) - which disappointingly gives no information at all on the problems that beekeepers were facing during the war.

Servicemen, returning injured from the front and unable to go back because of their disabilities, found that agricultural programmes which included beekeeping were available for them to follow as part of their rehabilitation programme. The setting up of the Land Settlement Association gave many returnees the opportunity of starting up in horticulture, a house and four acres of land being made available to successful applicants. At the end of the war, the journalist and beekeeper, John Charles Bee Mason, continued to work for the war office teaching beekeeping to soldiers in England and Germany.

In the United States courses were held in beekeeping especially for women at various educational establishments, but greater significance was given to training wounded soldiers to take up the craft once the war ended. In April

1919, The Federal Board for Vocational Education produced 'Monograph No. 37 Beekeeping - To the Disabled Soldiers, Sailors and Marines to aid them in choosing a vocation'. At the beginning of each of the books there is this note to ex-servicemen, which shows that the government was firmly committed to their rehabilitation and to placing them in jobs which they had helped to choose:

"Note to the Disabled Soldier, Sailor, or Marine.
As a disabled soldier. sailor, or marine you should remember always that the Office of the Surgeon General, War Department, and all of its employees, the Bureau of Medicine and Surgery, Navy Department, and all its employees, and the Federal Board for Vocational Education and all its employees, are mutually interested in your welfare solely. They have arranged a definite plan of cooperation to help you in every possible way. You can not afford to leave the hospital until the medical officers have done everything that they can for you to restore you to physical health and strength. Any other course will interfere with your vocational success later. Furthermore, you should by all means take advantage of the educational opportunities which the hospital has provided for you.

While you are making up your mind what line of work you want to follow you should take advantage of the opportunities to try yourself out in the different lines of activities which are provided at the hospital. When once you have made up your mind as to the employment you want to enter or the kind of training you want the Federal Board to give you after you leave the hospital, you should ask the vocational officers at the hospital to provide for you the kind of training which will advance you in the direction of the occupation which you expect to follow or for which you expect to be trained after you leave the hospital. You will find the educational officers at the hospital eager to render this service for you, and you should consult them early in your hospital career."

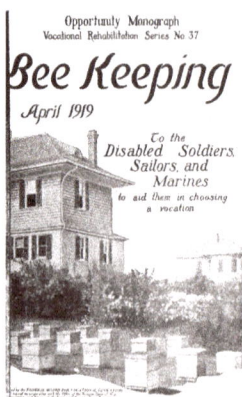

Opportunity Monograph
Vocational Rehabilitation Series No 37

Bee Keeping

April 1919

To the
Disabled Soldiers,
Sailors, and
Marines
to aid them in choosing
a vocation

Because of the scarcity of food during the war, ensuring that nothing was wasted was a major priority. Not only was food needed for the people at home, vast amounts of food were needed to be sent abroad to feed the large army. Propaganda posters alerted the population that food for the troops was needed if the war was to be won. Furthermore, there was the civilian population behind the front lines that needed feeding too, and towards the end of the war some 3 000 000 Belgians faced starvation.

Once the war was over, it took Britain a long time to recover from its effects. As regards beekeeping,

"We are saving you YOU save FOOD"

Well fed Soldiers
WILL WIN the WAR

KEEP it COMING
"We must not only feed our Soldiers at the front but the millions of women & children behind our lines"
Gen. John J. Pershing
WASTE NOTHING
UNITED STATES FOOD ADMINISTRATION

3,000,000 BELGIANS
ARE
DESTITUTE
IN
BELGIUM
They must not starve.
SUPPORT THE LOCAL FUND.

re-stocking of apiaries meant importing bees from abroad - when available. Britain learned that it could no longer rely on imports alone in the future and in 1919 the Forestry Commission was set up so that the country could produce a great part of its wood for the timber and paper industries. This undoubtedly changed the landscape with dark forests covering mountainsides or stands of timber taking the place of heather in the lowland wet heathlands. To ensure that there would always be a continuous supply of sugar, sugar beet was cultivated from 1920, thus changing the pattern of agriculture in the eastern part of the country.

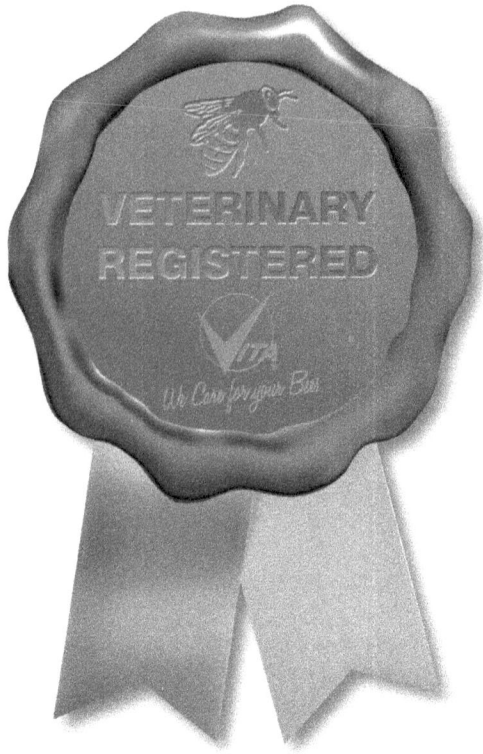

Vita prides itself on its commitment to the quality and efficacy of its products.

Choose our Veterinary Registered and approved products over those that don't make the grade to ensure the health and productivity of your Bees.

APIGUARD®   APISTAN®

www.vita-europe.com
Vita (Europe) Limited, Vita House, London Street,
Basingstoke, Hampshire, RG21 7PG, United Kingdom.

VITA
We Care for your Bees

DON'T WASTE BREAD !

SAVE TWO SLICES
EVERY DAY and
Defeat the 'U' Boat

THE FIRST WORLD WAR AND ITS EFFECTS ON SUPPLIES FOR THE BEEKEEPER AND HOUSEHOLDER

Don't You Know There's A War On?

Stuart Ching

Introduction

Britain continued to import food during the war. The main exporters to Britain were America and Canada. This meant that merchant ships had to cross the Atlantic Ocean. Up to 1916, these merchant ships could travel in relative safety. However, in 1917, the Germans introduced unrestricted submarine warfare and merchant ships were sunk with great frequency. This had a drastic impact on the nation's food supply and, with great losses in the Atlantic, food had to be rationed so that no one starved in Britain. In April 1916, Britain only had six weeks of wheat left and bread was a staple part of most diets. It was a bleak year for families - with food in short supply. Suddenly the war was brought home to most families. Food prices rose.

food
1 - buy it with thought
2 - cook it with care
3 - use less wheat & meat
4 - buy local foods
5 - serve just enough
6 - use what is left

don't waste it

The restrictions introduced by Defence Of the Realm Act (DORA) failed and reading the conditions one can understand why. It was a classic example of a 'knee-jerk' reaction by the authorities. The government then tried to introduce a voluntary code of rationing whereby people limited themselves to what they should eat. However, this did not work either.

August 1914

Early on in WW1 the problems of sugar supplies for everyone, not just beekeepers, became a problem. The BBKA had a system of supplying sugar to its members but this was sometimes under a strain. "We have received the following notice from our sugar merchants. 'Owing to the present state of the country, and the high prices ruling, we are only accepting orders at open prices, and subject to goods being obtainable. You may rest assured we will do our best to meet your requirements, and will not charge you higher prices than the conditions warrant. All previous quotations and list prices are withdrawn. These conditions also apply to all orders in hand unexecuted, and await your instructions.' "

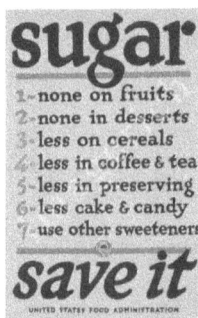

However, beekeepers thought they were in a special position as they considered that, when supplies of sugar diminished, they could use honey as a substitute. "The effect of the war is being felt by all, and even our industry has not escaped. Beekeepers should be careful to obtain the best price for their produce, as very shortly sugar will be unprocurable in this country, and honey will fill the gap at remunerative prices," wrote one correspondent to the BBJ. This was, of course, as long as the bees kept producing honey. We all know that the amount of honey produced each season varies enormously. In addition beekeepers were at the end of the season so no more honey would be available until the middle of the following year.

At first beekeepers themselves seemed a bit smug about their situation and needed the word of caution provided by the BBJ "The beekeeper with plenty of honey on his hands is lucky in that he can to a very great extent utilise it in the place of sugar, that has now gone up in price. Honey may be turned to account by the housewife in many ways besides just as a substitute for jam or butter, and is, in addition, one of the most wholesome foods that we possess. The price of sugar will have, too, a bearing upon beekeeping in another direction. Many stocks require to be sugar-fed to face the winter confidently, and it seems now pretty certain that but little of such feeding will be done.

The Editor of the BBJ, William Herrod, commented, "I have been much amused, upon asking several beekeepers how they had fared this year, to be told that they had got no surplus, and that the bees must take their chance this winter; they were not going to pay 4d. per lb. for sugar for them."

The 'Daily Mail' printed a useful tip and added a patriotic sentiment. "George Knowles of Codford St Mary informs us that the Roman fashion of sweetening food with honey has proved, on trial, excellent. Sugar is dear, and half a tablespoonful of honey will go as far as three tablespoonfuls of sugar. With sugar 100% up and likely to remain so, every ounce of honey in the country should be saved. Sugar will, no doubt, be dear and difficult to obtain, therefore it be well to practise a little self-denial in the way of sweets for ourselves rather than allow bees to become extinct."

1915

After a few months of war conditions, the reduction in sugar supplies from the continent started to strike home. Not without a tinge of jingoism the Westminster Gazette summarised, at length, the problem of reduction in sugar supplies. "Now that the importation of German and Austrian beet sugar has been suspended, the maintenance of an adequate supply of sugar at a reasonable price from other sources has already become a question of national importance. It has been urged that we should at once begin to cultivate sugar beet in large quantities and that refining factories, aided or owned by the State, should be established in England. Against these suggestions for the production of British beet sugar, I would ask for a thorough consideration on the part of the public and the Imperial Government as to whether it would not be better, both for the British consumer and for the trade of our great Empire, to turn our efforts to more fully exploiting the vast resources of the Empire to produce an increased supply of cane sugar to make up the deficiency.

It had long been recognised that beet sugar was unwholesome for bees, and amateurs were always warned against feeding their stocks with it; in this connection the British Beekeepers' Association had done splendid work in bringing pure cane sugars at a moderate cost to the notice of the public. Great Britain consumed 91 lbs. of sugar per head yearly, and though noted as one of the best sugar-producing countries of the world, she had been dependent on the inferior German product for her chief supply. It is, therefore, for the reasons above stated that I would urge, in this present crisis, the reorganisation and fuller development of our existing cane-sugar industries - putting within the reach of industries - putting within the reach of all, and at a moderate cost, the best sugar obtainable - rather than the experiment of making beet-sugar in England. The war has cut off the supply of inferior sugar; it gives us a unique opportunity to reinstate the best - our imperial cane sugar." The Gazette did not, at this time, join in the national call for beekeepers to produce more honey.

However, farmers were warned against the serious mistake of cultivating beet sugar, which is "a very inferior imitation of sugar" which is what was supplied recently from Germany, instead of the genuine cane sugar, which was said to be of intrinsic value from a medical point of view in strengthening the muscles generally, and the muscles of the heart in particular?

The Council of the BBKA had a long discussion on the matter of procuring sugar for the use of beekeepers, and eventually it was decided that under the present conditions of shortage for the general public it would be inadvisable to approach the Government in the matter. Beekeepers themselves were strongly advised to place their orders for sugar some time ahead of the period it is needed, either for syrup or candy making. As the orders were supplied in rotation by all sellers they would then obtain, if not all, a part of what they require.

1916

The country was well into the War by now and beekeepers had already had one season of wartime beekeeping and, like the rest of the population, were feeling the pinch of the comparatively hard times. How many beekeepers were tempted to take the last ounce of honey from the hives during the previous season then, because of the heavy price of sugar, neglected to see that their bees were well-supplied with good syrup?

The Board of Agriculture were encouraging the whole of the population to utilise every plot of land available to provide food for the people and this included the bee industry. "Honey was very useful, and was also more valuable as sugar was dearer. Honey was a staple that is sometimes classed as a luxury. It has a very real food value and at a time when sugar was so expensive, it should be more extensively used than ever before. Its production must be maintained." However, there did not seem to be a realisation that there is a connection between the production of honey and the necessity of feeding the bees with sugar through the winter. The production of honey is seasonal and so would not provide a regular substitute for sugar. Another problem in utilising land was the lack of labour to do so in the face of so much manpower being taken up by the armed struggle.

In an answer to a question posed by a reader, the Editor of the BBJ replied, "To give some idea as to the present price of honey is difficult. This we can only do in a general way, as prices will vary in different localities. At the present time good quality foreign honey is realising almost as much per ton as our home produced did before the war, and taking into account, also, the high price of sugar, and that from various causes our honey crop will only, generally speaking, be a moderate one, English honey should command a good price. In comparison to other commodities the price of English honey is still only moderate, and the scarcity of sugar is undoubtedly tending to stimulate the use of pure honey to a greater extent than before. Indeed, there is every reason for this, since honey is the most wholesome and healthful sweet that can be had."

The Government had by now taken over the control of the food supply of the country, and as there was such a shortage of sugar, it was suggested the time was right to get together the different resolutions passed by the various beekeeping associations all over the country, and get them sent to the proper authorities, showing the feeling of the beekeeping fraternity as to the necessity of legislation, in an endeavour to check the appalling loss of bees through disease, still going on all over the country.

The Royal Commission on the Sugar Supply arranged to grant a quantity of sugar not to exceed 50 tons in all for the purpose of feeding bees in the United Kingdom. Such sugar was only available in the form of bee candy. Its proper use was safeguarded by medication under the advice of the bee experts of the Board of Agriculture, and its manufacture and sale would be undertaken by Messrs. Jas. Pascall, Ltd., Blackfriars, to whom all applications, whether from persons in the trade or private individuals, were to be addressed.

It is quite possible the producers would not be able to supply odd pounds by post but, under this arrangement, it would be possible for any dealer in appliances, or even any grocer or confectioner in the most remote country village, to order a supply from Messrs. Pascall or their wholesale dealer. Syrup or plain sugar would not be supplied, and if syrup was needed it must be made from the candy. So, it is apparent that the Government did not trust the beekeepers not to turn their supplies to other (domestic?) uses!

1917

Reports appeared in newspapers on 3rd February 1917 that the government's food controller, Lord Devonport, wanted to avoid compulsory rationing but that there was a very urgent need to economise. People were asked to restrict themselves to 12 oz of sugar a week. Sugar was the first thing to be rationed more formally. In some parts of Britain sugar cards were issued by shops, for example by Co-operative Stores, giving their customers the right to a limited amount of sugar per week.

By now beekeepers were getting a little annoyed with the delay in the supply of the allocated candy. Pascall's would be able to supply "persons in the trade or private individuals," but an official of the Board of Agriculture stated that the method of distribution had not yet been decided upon; when it was, a notice would be published in the British Bee Journal.

Meanwhile beekeepers were contacting the BBJ Office asking for a supply of sugar. They were informed that plain sugar or syrup would not be supplied so when the latter was needed it must be made from the candy. The material used for "doctoring" the candy so that it would not be fit for domestic use was for that purpose only, and was not calculated to be a remedy for diseases.

Pascall's announced that they had now received from the Royal Commission on Sugar Supplies fifty tons of sugar for the purpose of supplying beekeepers with candy during the next few weeks. The candy, which had been stained pink to designate the purpose for which it was to be used, had been medicated with Bacterol, an antiseptic compound, which, whilst harmless to bees, rendered the sugar unfit for human consumption. It is said that the Germans with the same purpose in mind 'doctored' their candy with quinone – undetectable by bees but unpalatable to humans!

The candy could be obtained from at the following prices:
A five-pound box containing five one pound cakes at four shillings and seven pence, plus postage eight pence.
Four boxes at eighteen shillings, carriage forward.
Cases containing twenty-four boxes could be obtained at slightly lower prices, which would be sent on application.
The money had to be sent with the order.

One reader of the BBJ commented, "The sugar for candy making is now available, but if the supply set apart for that purpose is not used very shortly, the Food Controller will probably take what is left for other purposes. There is no prospect of the sugar supply being more plentiful for some time to come. It may become less and the price be higher. Although we all hope for a good season, there is no guarantee that we shall get one, and should the bees require feeding in the autumn it will probably be impossible to get sugar for the purpose. Any appliance dealer can secure the candy for his customers if he will do so, and, if it is ordered with other goods, a saving in carriage will be effected.
On the other hand, the increased supplies of imported honey ought to be more generally known. I wonder if the Sugar Commission is aware that this foreign honey is supplanting sugar, and in many cases surpassing it, especially the low grades of sugar the grocers supply at times. It astonishes me how very few people ever think that honey will take the place of sugar, in many cases with far better results; in fact, it is a sugar of the highest grade."
"Some hundreds of tons of sugar are being wasted annually in Kent alone, and everybody with a garden could do something to provide themselves with a supply of sugar." Such was the conclusion forced upon the minds of listeners to the admirable lecture delivered by Mr. W. Herrod-Hempsall, F.E.S., to the Kent Beekeepers' Association. He added, "It seems evident, therefore, that the effects of sugar restrictions can be greatly modified by the extension of intelligent bee-keeping, since there is no purpose for which sugar is used in which honey cannot be used equally well."
"With regard to the sugar scarcity, beekeepers are indebted to the efforts of the manager of the B.B.J., Mr. J. Herrod-Hempsall, in obtaining a grant of fifty tons for feeding bees. Again the Board of Agriculture went their own way in the matter of distribution, instead of a common-sense one, the result being that the candy sold costs about as much as honey. However, that is not the fault of Mr. J. Herrod-Hempsall, and no doubt it was the salvation of many stocks of those who could afford to pay the price.", reported the Secretary of

the Kent beekeepers,

To a bee
Yet how I envy you, a chartered rover
'Mid all the joys the countryside may yield,
'Tis literally yours to live in clover.
Garnering honey from the scented field.
Dame Nature never tasks your powers unduly;
You need but commandeer her endless stores.
Extracting sugar from my grocer truly
Is a tremendous task compared with yours.

"We have had a number of applications for sugar for bees. Will our readers please note that we are now quite unable to get or supply any sugar? We are sorry we cannot supply it, but it is only a waste of postage writing to us for it. Beekeeping has during the last few weeks received more notice than ever before. The shortage of sugar has emphasised the value of honey, and the B.B.K.A. has received hundreds of letters from would-be beekeepers, many of whom, we are afraid, will be no great acquisition to our ranks."

"We first tried honey for stewing plums last autumn, and we found it an excellent substitute for sugar." This from a housewife trying to make ends meet. "Seeing that it answered so well for the stewed fruits we tried it for jam making, and it answered the purpose splendidly. We could not get any sugar at the time for jam making, so we resorted to the use of honey only, and we made all our jams with it, and we are using the jam today, and it is splendid. If it had not been for the honey we should have been without jam some weeks now." This strategy was recommended by many newspapers of the time.

The Honorary Secretary of Stafford beekeepers was instructed to write a letter of thanks to Sir Walter Essex, M.P. for a Staffordshire constituency, for his services in obtaining from the Government the allotment of sugar for feeding bees, and expressing the hope that he would yet endeavour to get a reduction in the price, which at present was considered to be rather prohibitive.

"Many articles have already appeared in the BBJ which state how important it is to produce more honey, owing to the great scarcity of sugar caused by this terrible war; and so far as I can see the production of honey would be greatly increased if the boys and girls kept bees. The thing is, how is the knowledge of beekeeping going to be spread enough to make them do it. I suggest, and hope those who read this agree with the suggestion, that the experts of the different counties should give lantern lectures on bee-keeping in the public schools, if any, in their respective districts." Was this a case of class distinction from a reader of the BBJ?

One desperate beekeeper wrote to the BBJ, "Now my bees have about eaten up what honey they had got and I cannot get any sugar. Would the best treacle be any good, or what are other people doing under similar circumstances? With kind regards from an old beekeeper of years ago. C. Smith." to which the Editors of the magazine replied, "We have repeatedly replied to correspondents that golden syrup, treacle, glucose, etc., are not suitable for feeding bees."

"Sir W. Essex asked the President of the Board of Agriculture how much net profit has accrued on the bee candy made by Messrs. Pascall, to what benevolent fund has he directed its disposal, and whether under any suggestion as to classes of recipients?" The answer he received from Sir R. Winfrey: "The net profit which has been made on the disposal of the bee candy supplied by Messrs. Pascall, Limited, amounts to about £150. Among various suggestions for the disposal of the money under consideration, is one that the money should be handed to the Agricultural Benevolent Society for the benefit of beekeepers. I will communicate with the Hon. Member before the arrangements are completed."

Pascall's, the makers of the bee candy were forced to comment, "Since the making of bee candy was started, the price of sugar to us has gone up about three farthings a pound, but there has been no increase in the charge to the consumers, who are consequently getting the entire advantage in this respect. Also, unless there is a further rise the price will not be raised, although - in view of all the conditions existing at the present time - it is very close, and no further profit is anticipated. It will therefore be seen that there can be no suspicion of "profiteering in food" as regards the supply of it to the little creatures, who in turn provide us with a delicious and nutritious article of diet.

We pay more than you were aware of for sugar. We were doing the work practically for nothing in order to help the industry, and surely you do not desire our employees to also work for nothing. You might as well expect appliance manufacturers to supply hives at the cost of the wood only."

To this one reader replied, "I feel I should like to thank Messrs. Pascal for their letter of explanation, and also for the patriotic spirit in which they have dealt with the whole matter of making and supplying the candy, and, at the same time, point out where many beekeepers think Messrs. Pascal have been misdirected. I am a grocer, and, as I have been in the habit of making my own candy, I have generally found that the moisture added to the sugar covered the cost of making it into candy. Last year I could not get sufficient sugar even for my customers, as we are retailing the best cane lump sugar at 6d. per lb., and the candy was to cost 1s. per lb."

Each week the Spectator offered valuable general information. A correspondent raised an important question as to the price charged for sugar needed to feed bees at certain times of the year, and to the exorbitant price demanded for bee candy. "The Board of Agriculture might adopt a more helpful attitude and see that supplies are available for beekeepers at least on as favourable terms as sugar is permitted to reach brewers for the production of beer, which is not admitted by all authorities to be a food.

The whole of the 50 tons of sugar allotted for bee food early in the year had been used. It was pleasing to know that the Government had again recognised the requirements of beekeepers generally by allocating some further sugar for the purpose of manufacturing bee candy. It would be more pleasing still if they would allow it to be sold to beekeepers without its being first made into candy, but all efforts to get them to do this have so far failed, and beekeepers would be, and no doubt the majority were, thankful to get sugar for the bees at all."

1918

Sugar was often difficult to get. Whereas the weekly consumption of sugar was 1.49 lb in 1914, it fell to 0.93 lb in 1918. By the end of 1917 people began to fear that the country was running out of food. Panic buying led to shortages and so, in January 1918, the Ministry of Food decided to introduce rationing. Sugar was the first commodity to be rationed. The idea of rationing food was to guarantee supplies, not to reduce consumption. This was successful and official figures show that the intake of calories almost kept up to the pre-war level.

Rationing means
a fair share for all of us

Malnutrition was seen in poor communities. Food products were added to the list as the year progressed. Ration cards were issued and everyone had to register with a butcher and a grocer. One thing that was not apparently considered was that, apart from the issue of suggested amounts of food per household per week, (although the food may or may not have been available), people needed money to buy their allocation. Often this was a more severe system of rationing than the official one. The system was open to corruption, with those who had money buying up the rations of those who did not. The 'black market' in foodstuffs increased.

Herrod-Hempsall only mentions the war once in his booklet "Bee-keeping in War-time" which, produced in 1918 was unknowingly a bit late to help potential beekeepers during WW1.

"If there is plenty of honey in the brood combs for the bees to live on in the winter, there will be no need to feed. For this purpose there should be at least eight combs well filled and sealed over. If there is not this quantity then it will be necessary to feed with sugar syrup. It is well to remember that no food suits the bees so well as their own natural stores, therefore when removing the supers the beekeeper should not be too avaricious; if it is found that a stock has bred so well all the summer that practically all the stores gathered have been placed in the supers, then, under present war conditions, it will save much trouble and expense if one super-full is left on for food; if this is done the excluder must be removed or in the cold months it will prevent the bees from going through to the food."

Some readers of the BBJ were accused of being badly smitten with "Sugaritis." They were asked, "What has become of the honey your bees stored last season?" and "Would there have been all this 'weeping, wailing and gnashing of teeth' over lost stocks had the respective prices of honey and sugar been reversed last July?"

One beekeeper defended the method of distribution of sugar. "I cannot understand the rabid way the Government has been pitched into for arranging where we can get our excellent medicated bee candy - where can one get it cheaper and pure? I have fed my bees on it for two years, and my twenty-two hives are sound and well. I think the Government were perfectly right in preventing the sugar being used in any other way than for bees. I believe the scarcity of sugar will be a blessing in disguise to beekeepers in the British Isles, because they will have to feed more honey to the bees, which is their only true food. With the feeding of so much medicated syrup (sugar syrup) especially during the spring time, it is a wonder to me that the bees have any stomachs left."

With reference to the feeding problem, perhaps something might be done if beekeepers put their heads together. Others found, however, "The present system is very unsatisfactory, as the candy that is made by Messrs. Pascall is both dearer than sugar and for autumn feeding very troublesome to convert into syrup. We all appreciate, of course, the necessity of some means being taken to treat the sugar so that it cannot be diverted to human use after being purchased for bees. But surely this could be done by some means without making candy of it, and so save much time and labour and also fuel in boiling it down. I think, perhaps, it might be done by treatment with Izal. A very small quantity would make it quite unfit for human use, and at the same time serve as a disinfectant."

The notable beekeeper, ROB Manley, asked if anyone could suggest a workable plan - perhaps the B.B.K.A. would put it before the proper authorities. "It seems wrong altogether, when everything is so difficult, to waste time and money in converting a comparatively cheap food into a dear one, which in its turn, in most cases, requires again time, trouble and expense to reconvert it into a usable article. A quantity of sugar that would fulfil all needs for bees would be so small compared with the quantity required in the whole country as to be almost negligible."

The annual report of the Hampshire BKA stated that the season of 1917, although opening with a fair promise of success, could hardly be described as having been a good one for local beekeepers, and it was feared that, owing to a shortage of winter stores and difficulty in obtaining the necessary sugar for syrup feeding, considerable mortality had occurred. In fact, in three areas nearly 200 stocks were known to have succumbed from this cause. "Due to the scarcity of sugar everybody was rushing to honey as a substitute. Quite rightly; but many persons, knowing nothing of bees, or bee management, think all that is necessary is to place bees in the garden, and honey in galore will be secured. This seemed to be the reason for the large increase in people wanting to become beekeepers."

Pascall's began to feel the pressure of both the price of sugar and the demand from beekeepers. They stated that although the latest addition to the Sugar Tax had been in operation since April 23rd no increase in the charges for bee candy had yet been made. "Our bee-keeping customers had been given full advantage of the former prices for nearly three months. We were now compelled, however, to amend the charges, in order to cover the extra tax and increased cost of labour and packages."

1919
Beekeepers requiring a supply of sugar for feeding were told to apply at once to the Secretary of the Committee dealing with this matter in their respective counties for a registration form, which had to be filled in and returned to the source from which it was obtained. A certificate would then be issued entitling the holder to 10 lbs. of sugar per stock any time up to December 31st, 1919. This had to be presented to the Local Food Committee, which would issue the necessary coupons for the amount allocated from his grocer.

This was the situation beekeepers found themselves in at the end of the war. But, as we have seen since with all the subsequent wars, it takes a considerable time to get things back to 'normal'.

MAKING D0 WITH LESS
John Phipps

Not surprisingly, both during the war and afterwards - as it took a long time for nations to recover from its effects - householders were encouraged to make do with less on their plates and to use substitutes for commodities which were in short supply.

A good example of advice being given during these difficult times was published in *"Foods That Will Win The War And How To Cook Them"* (1918, Goudiss and Goudiss, USA). In fact, the changes people were advised to make, the book claims, would have a significant benefit on their health, as for too long people had consumed much more in the daily requirements of fats and sugars which were generally injurious to their well-being. Thus, from the Introduction:

"This book is planned to solve the housekeeper's problem. It shows how to substitute cereals and other grains for wheat, how to cut down the meat bill by the use of meat extension and meat substitute dishes which supply equivalent nutrition at much less cost; it shows the use of syrup and other products that save sugar, and it explains how to utilise all kinds of fats. It contains 47 recipes for the making of war breads; 64 recipes on low-cost meat dishes and meat substitutes; 54 recipes for sugarless desserts; menus for meatless and wheatless days, methods of purchasing - in all some two hundred ways of meeting present food conditions at minimum cost and without the sacrifice of nutrition.

Not only have its authors planned to help the woman in the home, conserve the family income, but to encourage those saving habits which must be acquired by this nation if we are to secure a permanent peace that will insure the world against another onslaught by the Prussian military powers.

A little bit of saving in food means a tremendous aggregate total, when 100,000,000 people are doing the saving. One wheatless meal a day would not mean hardship; there are always corn and other products to be used. Yet one wheatless meal a day in every family would mean a saving of 90,000,000 bushels of wheat, which totals 5,400,000,000 lbs. Two meatless days a week would mean a saving of 2,200,000 lbs. of meat per annum. One teaspoonful of sugar per person saved each day would insure a supply ample to take care of our soldiers and our Allies. These quantities mean but a small individual sacrifice, but when multiplied by our vast population they will immeasurably aid and encourage the men who are giving their lives to the noble cause of humanity on which our nation has embarked."

More specifically, the author writes:
"One ounce of sugar less per person, per day, is all our Government asks of us to meet the world sugar shortage, One ounce of sugar equals two scant

level tablespoonfuls and represents a saving that every man, woman and child should be able to make, Giving up soft drinks and the frosting on our cakes, the use of sugarless desserts and confections, careful measuring and thorough stirring of that which we place in our cups of tea and coffee, and the use of syrup, molasses or honey on our pancakes and fritters will more than effect this saving.

It seems but a small sacrifice, if sacrifice it can be called, when one recognises that cutting down sugar consumption will be most beneficial to national health, The United States is the largest consumer of sugar in the world. In 1916 Germany's consumption was 20 lbs. per person per year, Italy's 29 to 30 lbs., that of France 37, of England 40, while the United States averaged 85 lbs. This enormous consumption is due to the fact that we are a nation of candy-eaters. We spend annually $80,000,000 on confections. These are usually eaten between meals, causing digestive disturbances as well as unwarranted expense. Sweets are a food and should be eaten at the close of the meal, and if this custom is established during the war, not only will tons of sugar be available for our Allies, but the health of the nation improved.

The average daily consumption of sugar per person in this country is 5 ounces, and yet nutritional experts agree that not more than 3 ounces a day should be taken. The giving up of one ounce per day will, therefore, be of great value in reducing many prevalent American ailments. Flatulent dyspepsia, rheumatism, diabetes, and stomach acidity are only too frequently traced to an over-supply of sugar in our daily diet."

A COLLECTORS ITEM

❀

Rubber Collection Medal Featuring Bees

Britain's enemies during the 1914-1918 war had similar problems with shortages of food and other resources needed for the war effort. Rubber, needed for military use, particularly the aviation industry, was in great demand in Germany and the population was encouraged to scrounge around, look in every nook and cranny, for pieces which could be salvaged and recycled.

In return for their efforts, those people who were able to supply the authorities with useful quantities were awarded a medal in 1916 - as can be seen in the photos below. On the front of the medal is the inscription "Altgummi Sammlung" which means Scrap Rubber Collection. The design is made up of a Greek Cross with two ants and two bees looking into the corners where the lines of the cross intersect, and was created by George August Gaul, a Berlin based animal sculptor (1869 - 1921).

The reverse of the medal has an apt quotation from part of Goethe's "Chorus of Ants" from Faust, the whole verse of which is:

" As the gigantic ones
Have pushed it forward,
Ye pattering footed ones,
Swiftly arise ye!
Nimbly come in and out!
In such clefts as these,
Is every bit and crumb
Worthy possession.
The very best of all
Ye must discover,
Hasting most rapidly
Through every cranny.
Not idle must ye be,
Ye banded throngers;
In-gather ye the gold,
Heed not the mountain".

Note:
Ants - Herodotus - insects that collected gold dust
Bees - diligence and fervour; also Aryan symbol of the soul

OVERWINTERING HONEY BEES

John Phin

OVERWINTERING HONEY BEES

From the end of the 19th Century and for the following decades, it was most likely that beekeepers used calico rather than wooden quilts on top of the frames; indeed I still use them. I find them convenient because they can be peeled back carefully from one corner so that the condition of the colonies can be easily seen without removing them in their entirety - which is the case when wooden quilts are used.

However, one problem with them is that the bees cannot pass over from one frame to another to move on to other stores of honey. In a colony which has old combs, 'pop holes' in the combs would allow this movement to take place, but beekeepers usually replace these combs as part of their general management. In 'open' (mild) winters bees generally move across to new stores when there is a break in the cold weather; however, if there is a long, protracted spell of bad weather when conditions are severe, bees can easily become cut off from food supplies resulting in 'isolation' starving. It is not a rare occurrence for beekeepers to find dead colonies with plenty of stores, but sadly which were out of reach for them.

There are many unusual and interesting entries including the overwintering of bees in "A DICTIONARY of Practical Apiculture giving the correct meaning of nearly five hundred terms, according to the usage of the best writers, intended as a guide to uniformity of expression amongst BEE-KEEPERS. With numerous Illustrations, Notes and Practical Hints" by John Phin, author of "How to Use the Microscope", etc. and Editor of "The Young Scientist", published by E P Noll & Co, Philadelphia, 1884. (I do like the long book titles)

The first is a description and drawing of a simple piece of equipment which enables bees to move from one frame to another when calico quilts are used. This is then followed by the author's comments on winter passages in the combs:

Hill's Device - A contrivance by means of which the bees are enabled to pass *over* the combs, from one to the other, during vey cold weather. It is a substitute for winter passages (q.v.), but in our opinion is not equal to them. It has the advantage of not disfiguring combs, as the winter passages are said to do, though not to our eyes. As well speak of the combs as disfiguring the frames. Hill's device consists of a number of curved strips of wood nailed to a cross-piece as shown in the figure. When laid on the top of the frames it keeps the cover of the quilt up so as to allow the bees to pass under it. The cross piece is often made of iron, a great mistake, since the metal is too good a conductor of heat. Wood would be much better.

Hill's Device

Winter Passages - Passages made through the combs so that the bees can pass to the different combs without having to go under or around them. As the combs, where not covered by bees, are very cold - often frozen - in winter, any bee that attempts to crawl over them is lost, while if she could go through the combs, without leaving the cluster, she might be able to reach a supply of food and so sustain life. To enable her to do this, it is the practice of some of our best apiarists to cut holes in the combs about two or three inches below the top of the frames, and as the bees are apt to fill these holes up, many insert a tin thimble in them to keep them open. We object to tin, or any metallic substance amongst the bees in winter, and greatly prefer a wooden tube made by rolling a thick shaving round a roller, and securing it with very fine wire. The wood being a poor conductor of heat is generally better than tin. Such thimbles should be inserted in the foundation - thus saving the bees the labour of building comb which is to be afterwards cut out. At least four frames in every hive should have these thimbles. If they should come into extensive use they could be easily and cheaply turned out of some firm wood. The internal diameter need not be more than half an inch, and they should be quite thin. We have tried paper and pasteboard, but the bees gnaw them. We generally prefer two or three of these small holes to one large one. They should be at least three to four inches apart. Where tin tubes are used they should be heated and dipped in melted wax, so that the metal surface may be completely covered.

Hill's device (q.v.) is intended to answer the same purpose, but we do not think it quite as efficient. Perhaps *both* would be best.

Another strategy for overwintering bees in the colder parts of America was the use of the Chaff Hive, which John Phin also recommended for keeping the bees cool during hot summers:

Chaff Hive - A hive with double walls, the space between them being filled with chaff or some porous material which will prevent the passage of heat and will consequently keep the hives warm in winter and cool in summer.

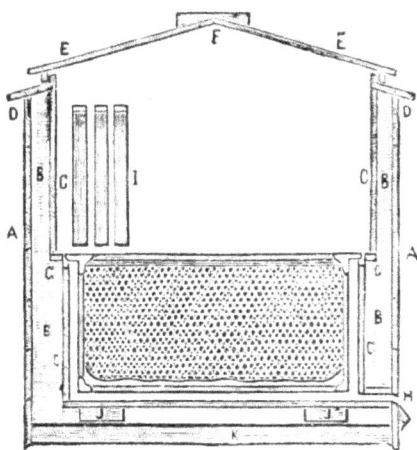

Section of Chaff Hive

A section of one of the most popular forms - the 'Simplicity" is shown in the accompanying engraving where A.A, are the outer sides; C.C, the inner walls, and B.B, chaff. H is the entrance, and E.E, a movable cover. It wil be seen that the hive is a two-story hive by construction, though of course it is used as a one-story hive in winter - the upper storey being filled with a bag made of cheap stuff and packed with chaff. A frame of comb is shown in the lower story with its end facing the entrance H, and three frames, I, are shown in the upper story and lying across it.*

"FALCON" CHAFF HIVE.

SIMPLICITY HIVE.

As Manufactured by W. T. Falconer, Jamestown, N. Y.

Chaff Hives

*Interestingly, Vasyl Priyatelenko, in Ukraine, uses this arrangement of frames. i.e. being at right angles to those below and above, in his hive which is based on how beekeepers construct their nest in a hollow tree. See: The Beekeeper's Quarterly No 113, September 2013.

BEEWAX

❦

ADULTERATION OF BEESWAX FOUNDATION

One of the reasons why some beekeepers today choose not to use commercially produced beeswax foundation is because of the fear that it might be contaminated with a variety of chemical pollutants. This concern was particularly highlighted, just a few years ago, when samples of wax comb taken from beehives in the United States revealed after analysis a cocktail of over a hundred pollutants. The chemical residues included those directly introduced into the hives from during the prophylactic treatment of colonies as well as from agricultural pesticides brought in from the surrounding environment by foraging honeybees.

As beekeepers usually recycle their wax and take it to their suppliers in exchange for sheets of foundation, it is most likely that the foundation will harbour chemical residues. This would certainly worry me if I was keeping bees in the United States, for the cancer causing paradichlorobenzine (PDB) is still being used for the control of wax moth. Indeed, I once wrote to the Editor of the American Bee Journal as I was appalled to see that one of his foremost columnists was recommending its use and, in reply to me, he couldn't understand what my objection was. At that time, in Greece, the two main wholesalers whose products flooded the supermarkets had to withdraw all their jars from the shelves because the PDB could even be detected in the aroma of the honey.

Spring airing of honey combs which have been treated with PDB over the winter months.

The increasing presence of beeswax being imported from China is also subject to some concern. In the past, honey and other products from that country have been found to be contaminated with chemicals or antibiotics, so there is not a great deal of confidence in the foundation produced from this source. In Greece - at least from my supplier - I can get two grades of beeswax: one is almost wholly from Chinese wax, the other containing a percentage of Chinese wax. This reminds me somewhat of a story I was told about milk after the explosion at Chornobyl. There was very little uncontaminated milk so it was mixed with contaminated milk, supposedly to lessen the radiation levels.

Of course, many beekeepers are willing to pay a premium price for 'organic' foundation, that is, when it is available. But there is, too, the growing trend of beekeepers who have a sustainable attitude to the craft and all the wax is produced within the apiary and recycled and, of course, without any chemicals being used when treating colonies for pests and diseases.

On the whole, apart from the possibility of chemical residues, it can be assumed that the wax which makes up the foundation is pure, i.e. it is made from beeswax and not other inferior and cheaper alternatives. However, this is not always the case. At a meeting I attended once near Kremenchuk, Ukraine, a presentation by two of my beekeeping friends describing the benefits of allowing the bees to build their own wax for cut comb honey production was suspended for twenty minutes when a major row broke out between the speakers, as well as in two sections of the audience, which nearly ended in a fist fight. I believe that the

There is a trend for beekeepers to let the bees build comb without the use of foundation.

comment was made that cut comb gave the customer honey in its purest form but not if foundation which might not be of the 'best quality' was used. Manufacturers and distributers of wax foundation objected strongly to the comment believing that accusations were being made at them and it took a lot of time before order was restored and the presentation be allowed to continue.

At a visit to a wax foundation factory a day or so later, the manager having showed me around his plant took me to his office and up to the window. On the windowsill, lying in full sun, were three strips of foundation from different sources and, judging by the time it took for each strip to begin to melt, it was possible in a very simple way to see the amount of adulteration in each sample.

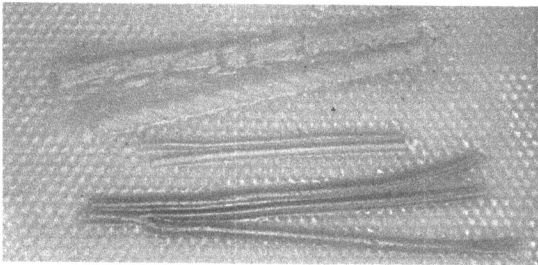

Foundation from different sources being tested on a hot, sunny windowsill.

At the beginning of the 20th century it was quite common, for a variety of reasons, for beeswax foundation to be adulterated with other types of wax. What follows here is a discourse on the subject by W B Webster (First Class Expert BBKA!) from his book **"The Book of Bee-Keeping"** Second Edition,1905, which was priced at one shilling, and in which he shows how beekeepers can test for themselves the purity of the wax they have bought.

Beeswax Foundation:
- is one of the most valuable adjuncts to modern bee-keeping; without it, honey in the form and quantity in which we see it would be unknown. The wax of which this is made must be absolutely pure beeswax. The price of such wax having, on account of the large demand for same, risen so considerably lately, and the price of foundation declined, has given rise to a considerable amount of adulteration with some descriptions of earth, wax, and fats. Beeswax melts at a temperature of 146 degrees Fahrenheit, while the melting point of other wax and fats is much lower. This adulterated foundation will not stand the internal heat of the hive, and so sags, or breaks down; in which case it is most likely to destroy all or most of the bees in the hive, drowning them in the honey which has been stored in the combs.

The bees will work out adulterated foundation almost as readily as pure. There is no doubt that bee-keepers are bringing this condition of things upon themselves. Many will have their foundation of a high colour in the case of stock, and perfectly white in that of super; they seem to forget the fact that perfectly white beeswax cannot possibly be obtained without subjecting ordinary coloured wax to a chemical change. The least deleterious method of doing this is by exposure to sunlight; but this entails so much labour, that it cannot be produced at the price charged for the foundation. The wax has to be made into thin sheets, exposed to the light for two days, then re-melted, made up again into thin sheets, again exposed to the light, and then melted into blocks. In consequence of the expense attending this process, other wax, or a mixture of chemically bleached beeswax with paraffin wax is used. Some super foundation we have tested had a strong odour of tallow. This was made from Cera japonica without a particle of beeswax in its composition, and was exhibited at a very large show as pure. As this wax is eaten with the honey from sections, beekeepers cannot be too particular that they obtain the genuine, unpolluted article, which, as foundation, is never white. In the case of stock foundation the same system occurs, bad-coloured wax being bleached, re-coloured, and scented, to make up for the loss of aroma in the bleaching process. Chlorine will bleach wax. but it makes it very brittle.

The following rough method of testing wax by its specific gravity will be found of service. Take a 6 oz. vial and half fill it with gin; then obtain a piece of beeswax that you know is pure, and, after kneading it into a ball about as large as a pea, drop it in the gin; it will sink to the bottom, as a stone; gradually add water to the gin, keeping it agitated while so doing, until the wax very slowly sinks. It is now ready. Take a portion of the foundation to be examined, and knead it into a ball of the same size as the first piece, taking great care that no air becomes embedded, drop this into the bottle, if it sinks in the same manner as the first piece put in, it is pure; but if it floats, it is not, as its specific gravity is lighter than pure wax. You must wait, before testing, until both pieces are of the same temperature.

Another and more satisfactory method is by testing its melting.point; this is performed by the following means: Having procured a thermometer that will register 200 degrees Fahrenheit, get a canned salmon or lobster tin, and remove both top and bottom; make some holes round the edge; in this tin place a small lamp. Next obtain an empty condensed milk tin, and half fill it with water; light the lamp, and place this on top. Now melt a small portion of the wax to be tested, and draw it into a capillary tube (a small glass tube, which can be obtained at any instrument maker's for a nominal sum); when cool, stop up each end, and tie it to the bulb of the thermometer; place them in the water, and wait until it registers about 135 degrees Fahrenheit, then watch the capillary tube. The very instant the wax turns transparent, and darkens, note what the thermometer registers, as that will be the melting-point. If pure, it ought not to register lower than 146 degrees Fahrenheit; you must not be particular to one degree with such rough appliances.

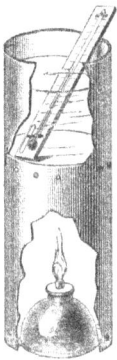

Apparatus for testing the temperature at which samples of wax melt.

There are several makes or patterns of foundation, Namely: Stock, super, natural base, flat-bottom, drone-size, and thick-wall. With the exception of flat-bottom, the septum or mid-rib of all is exactly the same shape as the bees naturally cushion their cells, but the flat-bottom, being simply a sheet of wax having the shape (hexagonal) of the edges of the comb-cells slightly raised on its surface, the septum is perfectly flat, i,e, quite unnatural. This latter make appears much thinner than the "natural base". The makers claim that the bees thin it out more perfectly than they do the "natural base". This is not so. The foundation looks better for sale; that is the only advantage and a very questionable one to the bee-keeper. Super foundation is made of better-coloured wax, also thinner, than stock, as it is used only in the sectional supers, while stock is kept entirely for the body-box and shallow frame supers. The foundation we advise for use is the "natural base", both for the brood-box and the supers. Sectional supers should be fitted up with thinner foundation, that is, super foundation, which on no account should be made of white wax (usually adulterated), but of yellow, with plenty of honey-like aroma to it. When held up to the light it should be very clear; if it has a granular appearance, after warming, there is almost sure to be an adulterant present. Foundation over twelve months old is exceedingly brittle and hard. This can easily be remedied by thoroughly warming it either before a fire, or better, by immersing it in water heated to a temperature of from 100 degrees to 105 degrees Fahrenheit.

MAKE YOUR OWN FLAT FOUNDATION

In the section above about beekeepers who wanted to keep their bees in a sustainable way, I mentioned that often the bees are allowed to build comb without the use of any foundation. However, many beekeepers who use top bars to support the combs, may well provide them with strips of wax, recycled from the cappings of combs during honey processing. All beekeepers know that making the bees build new combs each year for honey storage may adversely affect the total amount of honey harvested so, when plentiful wax is available in the apiary, attempts are made to make their own foundation. At one time some simple foundation kits were able with a silicon former, so that the beeswax produced in the apiary could be embossed with the hexagonal patterns - as one finds with commercial foundation. However, is the embossing really important for the bees to help them build their combs? This question is simple to answer - without embossed foundation, the bees would draw out a mixture of worker and drone cells - just as they would build any free comb within the hive.

David Dawson, the Canadian correspondent for The Beekeepers Quarterly, at one time wrote about his experiences of producing flat wax. The process is very simple and can easily be adopted by any beekeeper:

1. Requirements
    (a) - a supply of clean wax
    (b) - a means to melt the wax
    (c) - thin wooden boards
    (d) - cold water to wet boards
    (e) - exact size trimming board
    (f) - embedding tool (if wax is to be wired)

2. Melting Pot with Wax

3. Thin Wooden Boards a Little Bigger than the Frames

4. Soak the Boards in Cold Water

5. Dip the Board in Wax

6. Turn the Board Over and Dip Again - two coats are best

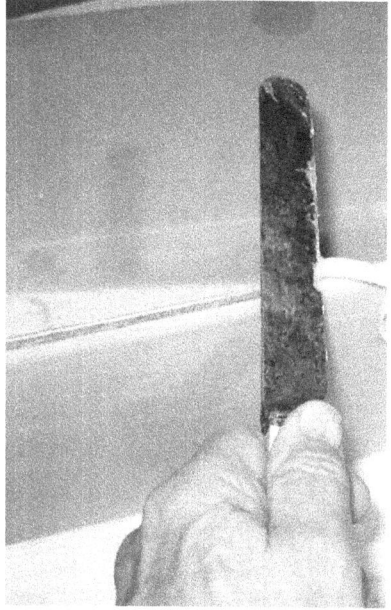

7. Trim the Edges whilst the Wax is on the Board

8. Peel Off the Wax Sheet

9. Pattern of Exact Size for Frames Being Used

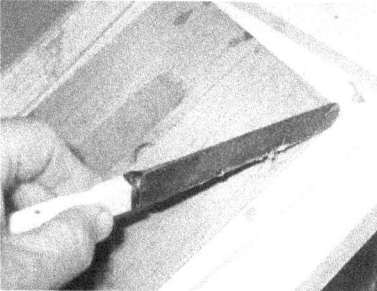

10. Trim Sheet to Exact Size

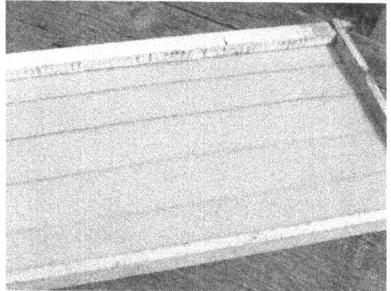

11. Put Wax in Frame and Melt Wax to Wires

12. Heat Grooved Wheel

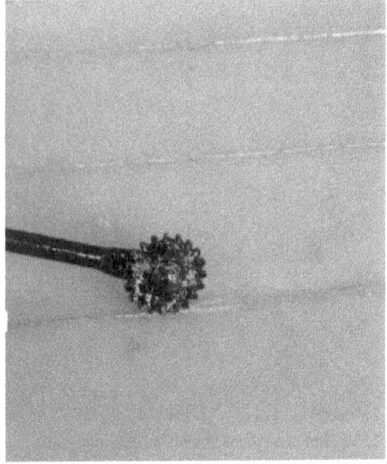
13. Roll Hot Wheel Along the Wire to Melt the Wax

14. Cut Holes in the Wax so that Bees Can Work from Both Sides

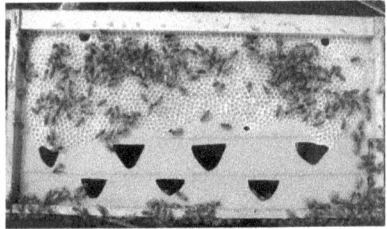
15. After Three Days ...

15. After Three Days ...

David believes that a swarm would draw out such sheets of wax very quickly and that if the flat sheets were put into an expanding brood nest, then worker size cells would be built.

This was the first experiment carried out on this technique and he saw that the system had good potential for areas in the world where embossed foundation is not easily available.

ORIENTATION OF FOUNDATION

Does it matter if foundation is placed in frames at 90 degrees to its norm? We are used to the cells of honey combs having the vertical sides parallel with each other. I always believed that this was because bees use the forces of gravity during comb construction. But does it really matter to a colony's performance?

Vasyl Priyatelenko in Ukraine uses frames of two sizes in his hive. In the lower and upper chambers, the frames are broad and shallow, whilst the middle box has very tall frames, which are put at right angles to those in the other boxes. However, in the former frames the foundation is placed in the normal way, whilst in the deeper frames the same foundation is used but rotated by 90 degrees. So, what do the bees do?

As can be seen in the second photograph, the cells are built according to the design on the foundation - i.e., the parallel walls of the cells are, as you would expect, almost horizontal. The bees seem to have no problem rearing brood or storing honey when the cells are constructed in this unconventional way.

The same foundation used both horizontally and vertically in the frames.

Completed comb with brood and honey with parallel sides almost horizontal. (Photos - Victor V

39

40

TELLING THE BEES

BILL CLARK

❁

BEEKEEPERS WANT TO BREAK WITH ALL TRADITIONS?

Having regard to previous sorties with beekeeping news to the press, I well know that to guarantee an airing it needs to be sensational! To gain at least a seventy-five percent chance of a column inch or two, an appealing story or photo will do the trick. But insistence on a straight piece of beekeeping news will bring the odds down to less than evens that you will be ignored. With this in mind, when I wanted to make sure that the public at large did not miss the passing of late Committee member and personal friend, Harold Schurr, I reasoned that the old tradition of dressing the hives and telling the bees of their Master's death would come well within that second category.

Harold was a very private, but scientifically minded man, and a quite well-known Biological Agronomist, and I deemed it only sensible to have a word with Jenny, his widow, first. Her answer was, for want of a better word in such circumstances, enthusiastic, and she believed that Harold himself would have been appreciative of the thought.

I hastily looked through my books, for I wanted to do the job correctly, but not a mention could I find in any of the recent, so-called encyclopaedias. Book after likely book piled up on the floor, until I picked up the 1917 ABC-XYZ of Beekeeping, and on pages 121 and 122 Mr. Root speaks of beekeepers in the Southern States still using box and log hives, and of others in the isolated, mountain regions, still clinging to Anglo Saxon beliefs. Among the examples I found: telling the bees of a death in the family etc, etc, and he then goes on about the urgent need to educate these poor people. As a matter of interest I looked in the 1947 ABC-XYZ, and found the section much the same, except that all mention of these folk practices had now been expunged! I moved along the shelves, cross with myself, that I could not find what must be somewhere, then I picked up my Ex-library copies of Hempsall's 'Beekeeping New and Old' and there, at last, on page 1786: TELLING THE BEES. Not only full descriptions of the various verses, and modes of carrying it out on the next four pages, but a splendid photo of a lady (possibly a genuine daughter) leaning over suitably bedecked skeps, telling of a death in the family. I can't help wondering now if it was not those very pages that decided some County Librarian that these books were now too out-of-date to keep. (Thank goodness, I say!!)

By mid-day on Saturday the hives were be-ribboned and for good measure I had placed a draped skep at the front. The Cambridge Evening News (CEN) photographer arrived, and bee-allergic Jenny stood near to see and listen. I did in fact find the ceremony quite moving, despite being asked by the photographer to repeat, either the tapping with the key or the placing of the ribbon, about twenty times.

On Monday morning the phone rang at about 8.30 am. A reporter at the other end explained that she had 'these quite wonderful photos' on her desk and could I tell her the full story behind them. "We must get it in today's edition" By late afternoon, visitors to the Country Park and Nature Reserve that I run were calling out as they passed, "Like your photo in the paper!" That evening ours arrived. 'Bingo' Not only in colour! Not only a quarter page! But bang on the front page! At 8.30 that evening the phone rang again. It was a fellow beekeeper, "I have just picked up the Evening News. How on earth did you get yourself involved in this rubbish on the front page?"

Over the course of the next few days either by 'phone or word of mouth, people wanted to know about bees in the garden, bees in the walls, how to start beekeeping. Two schools required visits, and one business man even suggested paying for my services - to visit his garden and advise as to suitability, and give him, one to one, education on keeping bees and provide all the equipment, money no object! And two more indignant beekeepers got in touch; one said that until now he had always thought that "I was quite a sensible person" and the other wanted to know "Do you really believe in this stupid clap-trap?"

At the end of the month, the CEN ran the photo again on their page of stories and photos of the month, and yet another spate of bee enquiries came in. One visitor approached me with "Hello Mr. Clark, I had no idea that you were such an expert on bees!"

So there we are, I now have the proof that we beekeepers have a different mentality to other folk. Let's be proud of it, show our solidarity. From now on let's boycott all those wretched agricultural shows that make a spectacle of farmers using great clod-hopping cart horses, and traction engines. Whilst we are at the Honey Show, stop the rest of the family from seeing the Changing of the Guard, Black Rod, Beef Eaters and all such clap-trap (I take it we are all in agreement with keeping the Monarchy!!). And above all, just to prove we really are in the twentieth century, never, ever, again sit in front of a log fire with a glass of mead in our hands.

TELLING THE BEES

It was the practice, and indeed it is still followed in some parts of the British Isles, to tell the bees of weddings, christenings, deaths and other important events in connection with the bee-keeper's household. In most cases the use of the key of the house door is considered as essential in the ceremony as the wand of a magician in the exhibition of his art. It was imagined that if the bees were not apprised of these events they would either abscond or die. If neither of these incidents occurred, then some member of the family would either die or meet with grave misfortune.

The following is a self-explanatory Irish ditty:

> A maiden in her glory, Upon her wedding day,
> Must tell the bees her story,
> Or else they'll fly away.

The mode of telling the bees of a death varied only slightly in different parts of the country, therefore a general description will suffice. Even in the depth of winter, as well as in the summer, it was considered necessary for a near relative to "wake the bees" by taking the key of the house door and solemnly tapping each hive three times, and repeating at each stroke, "Bees, bees, master (mistress or other member of the family, as the case might be) is dead." If it was the owner who had departed, the following was added:- "You must now work for (giving the name of the new master)".

It was usual also to put the bees into mourning, either by attaching a bow of black crepe to each occupied hive, or by fixing a stake in the centre of the apiary and draping it with the same material. In addition, it was deemed necessary to lift the hives and place them back on the original stand having the doorway facing in the opposite direction. In Devonshire it was said to be imperative that the ceremony of turning the hives was to be performed on the day of the funeral of the owner, and at just the moment the corpse was being carried out of the house.

A variation of the above was to drape the hive, and at night the bees were woke up "by sharply rapping on it with the knuckles, at the same time informing the inmates of the event" the humming of the bees caused through alarm at being "woke up" in such a manner was considered to be their response to the communication.

The writer has known crepe to be pinned to the hives with the idea that by so doing health would be restored to a member of the family lying seriously ill.

In some parts an opposite attitude to telling the bees was adopted, viz., should the widow of a male owner find that the bees were so unsympathetic to their loss as to continue to live, the punishment meted out for their lack of respect was to turn all the hives upside-down and so expose them to the elements until they died.

In the 'Living Librairie,' translated by John Molle in 1621, p. 283, appears the following :-

"Who would believe without superstition (if experience did not make it credible) that most commonly all the bees die in their hives if the master or mistress of the house chance to die, except the hives be presently removed into some other place? And yet I know this hath happened to folke no way stained with superstition."

In some districts it used to be the custom to lift the hives occupied by bees whenever a funeral passed the apiary. This was considered important to the well-being of the bees.

In Yorkshire not only were the bees put into mourning, but when the funeral feast, or as it is called in that county 'the arval,' was held after the mourners returned from the burial, one of the family had to give the bees a taste of everything partaken of by the guests, such as wine, cake, cheese, ale, and even tobacco.

The following poem, taken from 'Songs of The Ridings,' by F .W. Moorman, describes the incident of telling the bees of the passing of their master:-

"Whist Laatle bees, sad tidings I bear,
Bees, bees murmurin' low'
Cauld i' his grave ligs your maister dear,
Bees, bees, murmurin' low;
Nea mair he'll ride t' soond 0' t' horn,
Nea mair he'll fettle his sickle for t' com,
Nea mair he'll coom to your skep of a morn,
Bees, bees, murmurin' low.

Muther sits crying' i' t' ingle nook,
Bees, bees, murmurin' low;
Parson's anet her wi' t' Holy Book,
Bees, bees, murmurin' low;
T' mourners are coom, an' t' arval is spread,
Cakes fresh frae t' yoon, an' fine havver-bread,
But toom is t' seat at t' table-head,
Bees, bees, murmurin' low.

Look, conny bees, I's winndin' black crepe,
Bees, bees, murmurin' low;
Slowly an' sadly your skep I mun drape,
Bees, bees, murmurin' low;
Else you will sicken an dwine reet away,
Heart-brokken bees, now your maister is clay
Or, mebbe, you'll leave us wi' t' dawn 0' t' day,
Bees, bees, murmurin' low.

Sitha! I bring you your share 0' our feast.
Bees, bees murmurin' low;
Cakes an yal an; wine you mun taste,
Bees, bees, murmurin' low;
Gie some to t' queen on her gowlden throne,
Ther's foison' to feed both worker and drone
Oh! dean'it let us fend for ourselves alone
Bees, bees, murmurin' low.

Glossary:
yoon - oven; *toom* - empty; *conny* - darling; *dwine* - waste; *yal* - ale;
foison - plenty.

And again, J.G. Whittier in his "Telling the Bees" wrote:

Just the same as, a month before,
The house and the trees,
The barn's brow gable, the vine by the door,
Nothing changed but the hives of bees

Before them under the garden wall
Forward and back
Went drearily singing the chore-girl small
Draping each hive with a shred of black

Trembling, I listened: the summer sun
Had the chill of snow;
And the song she was singing ever since
In my ear sounds on :-
"Stay at home pretty bees,flying not hence!
Mistress Mary is dead and gone!"

Glossary:
chore girl - servant girl

- From Beekeeping New & Old Described with Pen and Camera, (1937) Vol. II by W. Herrod-Hempsall, F.R.E.S .

EDITOR: The ceremony of "Telling the Bees" varied in different parts of the country. Mary Webb in "Gone to Earth" comments on the custom in her native Shropshire: *"Hazel Underwood's father was a coffin maker and a beekeeper and whenever someone ordered a coffin, Hazel would go and 'Tell the Bees' who had died. Abel told his daughter it wasn't necessary – the bees only needed to be told when someone in their own house was dead – a simple ceremony he carried out himself when his wife died; 'Maray's jead,' he solemnly announced to the bees on that 'vivid June day.' "*
From A Beekeeper's Progress, by John Phipps, Merlin Unwin, 2013.

FROM MAQS TO QSMOS
- BOOKS FOR 2014

John Kinross

The time has come round again for the Taranov Board to meet to choose the books of the year, 2014. There are some readers new to the Annual who don't know Professor Dripitoff (a Harley motorbike enthusiast), Ina Strainer (who seems to do all her beekeeping in the back of her Morris Traveller), and myself - whose last active beekeeping was in St Helena a few years ago. We meet in the local pub to shout, above music, suggestions on the books we like.

Ina gets in first to say her local bookshop closed just before last Christmas, but not before she had found **Dorothy Hartley's "Food in England"** in paperback at £19.50 (Piatkus Press). She says all cooks need it in the kitchen. For beekeepers there are five pages with countrywoman Dorothy giving the value of a load of hay in their paddock in 1945 as worth £7, a silver serving spoon, priced by weight, worth £5, and a swarm of bees during the sugar shortage was worth between £5 and £10 - certainly "not worth a fly" as the old sayings incorrectly state. This book was published in 1954 and is well worth re-discovering.

The Professor likes large coffee-table books to impress his friends and **Chris O'Toole** has not disappointed us with **"Bees - a Natural History"** (Firefly £30) which has some great colour photos, some old prints, and covers Solitary Bees, Folklore, Squash Bees (Ina thinks they prefer lemon squash to sugar and water) and any other kinds of bees you can think of.

I have just found **Richard Jones'** book of **"The Folk Art of Slovenian Hive Fronts"** (IBRA £14.50) with colour pictures of scenes from country life painted on hive fronts. 'Sawing the Hag' is a bit gruesome but some of the religious themed pictures, particularly the Magi, are excellent.

David Heaf, the chief protagonist for Warre hives of the top-bar variety has now written a manual on **"Natural Beekeeping with the Warre Hive"** (NBB £14). He recommends some special tools - a cheese cutter (Ina says one of her old boy friends had a cheese cutter hairstyle when he was a National Serviceman), a brush with non-synthetic bristles and some home-made hive tools. The Professor uses his Harley spanner to unstick his hives.

The editor of this Annual and The Beekeepers Quarterly has recently taken to top-bar hives. **John Phipps'** autobiography with keeping bees in Greece and other places **"A Beekeeper's Progress"** (Unwin £14.95), is a handsome publication with small paragraphs, lots of black and white illustrations and reads very well. Perhaps an Index would have been a good idea but, as it is not a text book, this is not essential, though the Chernobyl trip by the author could have been mentioned there (It was! Ed).

Reprints, not on varroa, are still appearing. One of the more unusual books to have just been re-issued in a de luxe edition is **"The General Apiarian"** by **Jacob Isaac**, 1803 (BBNO £90, or boxed edition £130). Jacob started the Western Apiary Society which met in a pub near Exeter Cathedral. He recommends the 'Preserver and Remunerator Hive' and strongly attacks those who kill off their bees at the end of the season. Alas, his society didn't last after his death in 1819. This is the problem with small societies when it is one man or woman who does all the work.

Bumblebees are not normally mentioned in this annual, but Somerset BKA have **Dave Goulson** as their main speaker in their 2014 lecture day. Goulson is the man who started the Bumblebee Conservation Trust in 2006 at Sterling University. His book is about the introduction of the short-haired bumblebee back to England - Kent, to be precise, with much on his foreign travels, but also useful items like the Latin names of bumblebees with their English equivalent. There are no illustrations and the price reflects this (Cape £16.99).

Finally from the USA comes **"Natural Beekeeping - Organic Approaches to Modern Apiculture"** (Chelsea Green, USA £25.99) by **Ross Conrad**, a member of Vermont BKA. He uses sugar and garlic powder to control varroa and MAQS (Mite Away Quick Strips) for his hives. The Professor remembers QSMO at his local pier show when he was younger - 'Quick Strips - Men

Only' it stands for, it seems. Conrad's book is a large paperback with a glossary that is not always relevant in the UK. Ina thinks a 'Birthing Cell' is the maternity ward of her local hospital and the Professor thinks that 'Bottom Bar' is where he can enjoy his cider and pipe in the local pub cellar.

Finally, the beginners' book that has taken some time to be accepted now starts coming into general use. **Ivor Davis** of the BBKA has written **"The BBKA Guide to Beekeeping"** (Bloomsbury £16.99) with help from his designer and beekeeping friend Roger Cullum-Kenyon. The book has lots of colour plates and sound advice; a book for every beginner.

As this article is written at Christmas Cracker time the best joke we have heard is "Who is Santa's No 2?"

The answer - "Subsidary Claus(e).

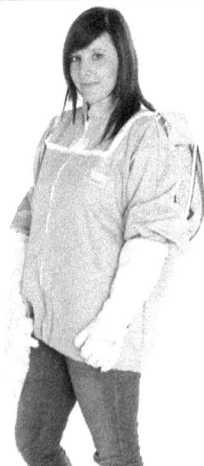

SMALL HIVE BEETLE - TO CLOSE FOR COMFORT

Small Hive Beetle (*Aethina tumidea*)

The news that Small Hive Beetle (*Aethina tumida*) was discovered in Italy in September is undoubtedly bad news for all beekeepers in Europe. Previously, the pest, which can devastate honeybee colonies, was found in the larger continents of Africa (where it originated), the United States, and Australia.

The confirmation of the outbreak was made by the Italian National Reference Centre for Beekeeping on September 11th, following analysis of samples taken from bait traps, belonging to the Agricultural University of Reggio Calabria, in the vicinity of the port city of Gioia Tauro, on the 5th September.

Following its discovery, urgent measures were said to be taken to determine the extent of the outbreak which included sampling colonies within twenty kilometres of the original sites well as tracing colonies which had been moved in and out of the area - either as part of beekeepers' migrations to other areas or the selling and purchasing of colonies.

In areas where the beetle is found, other controls are in place, including the ploughing of land around the apiaries the treatment of the soil with a pesticide drench.

The BBKA, on receiving the news, commented on their website:

"Since 2011, there has been a substantial level of imports of package bees and queens from Italy into the UK. The NBU is arranging for further inspection of colonies belonging to these beekeepers.

Mike Brown and Gay Marris are attending the EURL Honey Bee meeting in France next week where the Small hive beetle will be discussed extensively. For more information about this exotic pest and the things beekeepers should do are illustrated in the NBU advisory leaflet 'The Small hive beetle'. Please see the NBU website for more details: https://secure.fera.defra.gov.uk/beebase/"

Checking Colonies for the Presence of Small Hive Beetle
(from DEFRA's Small Hive Beetle Advisory Leaflet)

The following method is useful for the detection of all life stages.

1. Method: Scanning combs and boxes
It is important to use a torch when examining your colony(ies) for Small hive beetles - particularly in any apiary with tree cover, which can make the site dark even on a sunny day.

Carefully remove the hive roof and check for adult beetles running around under the lid. Then place the roof upside down next to the hive. Remove the supers and upper brood chamber (in double brood chamber colonies), and place them on the upturned roof for a few minutes. Place the crown board on top. A few minutes later, lift the boxes out of the way and scan for beetles on the inner surface of the upturned roof. When hives are opened adult beetles quickly scuttle away from the light, so look for adult beetles moving inside the hive, running across the combs, crown boards and the hive floor.

In warm weather, adult beetles will mostly be on the hive floor; in colder weather they hide themselves in the bee cluster for warmth. Look for clusters of eggs (two-thirds the size of bee eggs) in irregular masses usually in cracks and crevices in the hive. Look for larvae in the combs or on the bottom board.

Remove the combs one at a time from each box, and carefully examine each of them for evidence of adult beetles and damage caused by the larvae. Although they may at first glance look like wax moth, beetle larvae can easily be distinguished after close examination. Note that it is very difficult to detect low numbers of Small hive beetles in hives, so regular inspection of colonies in apiaries is essential for early detection.

Method: Using corrugated hive floor inserts

A simple detection method, using either cardboard or corrugated plastic hive-floor inserts, has been used successfully for detecting the Small hive beetle. This exploits the beetle's tendency to seek dark crevices in which to hide. Corrugated plastic is longer lasting and can be obtained directly from appliance manufacturers or made up by the beekeeper. If making your own cardboard insert (Figure 25), remove the paper from one side to expose the corrugations. The upper side of the trap then needs to be 'faced' with plastic tape to prevent the bees from chewing it up and throwing it out of their hive. Place your trap, corrugated side down, on the bottom board towards the rear of the hive. Regularly examine the debris under your insert(s) for evidence of adult beetles or eggs in crevices on the hive floor. Whatever type of corrugated insert you use, it is really important that immediately upon removal, you put the trap into a clear plastic bag when examining it - otherwise any beetles will easily escape.

Control of Small Hive Beetle

It is has been found, in practice, that very few - if any -propriety chemical treatments are effective for protecting colonies from small hive beetles. However, various traps, of various degrees of complexity, are available, but they tend to be very expensive.

Ann Harman, in the USA, suggests the following ways of dealing with the pests:

Oil of Wintergreen

1. In the US we can get from pubs and bars, cheap pressed paper coasters - i.e. beer mats. These are quite absorbent and about 8 cm diameter. If beer mats are unavailable then use some sort of pressed absorbent paper of the same size.

2. Next, obtain Oil of Wintergreen - (exactly that -- everyone seems to be into 'essential' oils today so it should not be a problem to find a source).

3. Put three, (yes, 3!) drops of wintergreen oil on the coaster. Place one on the top bars of every hive.

This chases the SHB out of the hive so if you do not put one on every hive the beetle will just find a hive without odor and move in.

Yes, the bees will eventually chew up and throw out bits of coaster. Get more.

Yes, you may renew wintergreen when the coaster stops being smelly.

No, it does not kill SHB - it chases them out and keeps them out.

If you are stupid and put more than 3 drops on you will chase your bees out, too.

The source of this advice is an excellent migratory pollinator, running about 4000 hives in western Kentucky. He pollinates crops like watermelons that grow in the loose sandy soil in that area. SHB LOVE loose, damp, sandy soil to pupate!

SHB has great difficulty in bad soil like heavy clay. Also they cannot stand low humidity -- can't pupate. Some of the big honey producers run dehumidifiers in their big honey houses.

Sure is cheaper than all sorts of new equipment!

B. Chickens.

A fenced apiary, with hives on goods stands, can be kept free from small hive beetles if chickens are allowed to roam around. The pupal stage of the beetle lasts from four to six weeks and takes place in the soil - giving chickens a long time to forage for this excellent source of protein.

Chickens are excellent for controlling small hive beetles as they can consume many of the earthbound pupae.

WORD SEARCH

THE GRID BELOW CONTAINS THE NAMES OF THE GENERA OF FIFTY IMPORTANT BEE PLANTS. THEIR NAMES MAY BE FOUND BY SEARCHING THE GRID HORIZONTALLY, VERTICALLY AND DIAGONALLY, BOTH BACKWARDS AND FORWARDS..

C	M	A	C	I	V	A	T	A	O	I	M	G	S	U	M	Y	H	T	U
R	Z	C	X	H	S	S	I	E	T	E	A	A	R	P	B	U	X	U	S
A	T	J	R	U	Z	E	N	N	E	L	N	V	H	Q	B	Z	X	N	S
T	S	A	L	E	C	Y	E	E	A	P	L	O	R	O	B	I	N	I	A
A	C	A	R	N	V	Y	C	N	I	T	I	M	T	R	N	X	X	B	F
E	M	O	A	A	O	A	T	A	L	T	S	L	G	H	N	I	U	Z	M
G	E	Z	R	X	X	H	P	I	L	I	A	A	O	I	E	S	A	U	H
U	B	O	D	Y	U	A	S	A	S	L	W	P	C	B	U	R	I	B	C
S	R	R	I	S	L	G	C	Z	P	U	U	C	M	H	O	L	A	Q	M
X	A	I	O	T	L	U	R	U	B	U	S	N	T	I	O	L	A	R	B
I	S	G	K	H	G	J	S	Q	M	H	P	N	A	F	A	L	I	O	F
L	S	A	S	C	R	O	P	H	U	L	A	R	I	A	U	B	R	U	A
A	I	N	S	E	N	E	C	I	O	I	F	R	S	D	E	A	H	M	M
S	C	U	C	R	O	C	U	S	L	X	T	S	N	S	G	J	Q	F	V
S	A	M	J	S	D	N	A	E	C	V	I	A	V	O	N	L	C	H	U
A	E	A	H	T	L	A	H	O	Z	L	V	A	I	L	E	C	A	H	P
Q	S	O	E	R	I	C	A	N	E	A	F	A	R	B	U	T	U	S	R
S	I	S	P	A	N	I	S	M	L	F	A	G	O	P	Y	R	U	M	E
O	V	C	J	A	H	T	N	E	M	U	R	T	S	U	G	I	L	Q	C
W	X	O	N	O	B	R	Y	C	H	I	S	Q	R	U	S	C	F	O	A

ANSWERS ON PAGE 200

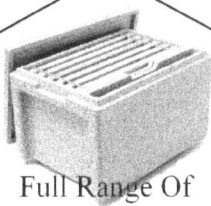

15

DIARY & CALENDAR

- PART II -

*SR (SUNRISE) SS (SUNSET) FOR LONDON UK.

JANUARY

January blossoms fill no man's cellar.

TAYLOR'S UP-TO-DATE 20th CENTURY HIVE.
Fitted with swarm-preventing chamber, containing shallow bars, brood nest with 10 standard bars, porch and entrances 6 in. lift containing crate of 21 1-lb. sections, 6-in. lift for tiering, quilts, roof and legs, etc.
18/- complete.

CHAIN-GEAR EXTRACTOR, 28/-

No. 17—With Cowan Cog-Gear, 26/-

No. 22—**NATURAL BASED OR ROOT FOUNDATION.**

GIRDER WINTERING DEVICE, 4d. each, 3/6 per Dozen.

No. 51—In 3-gross Cases, 18/-
No. 51—1-lb. Screw-top Bottle, 20/- per gross.

1st Quality Brood 2/-

1 to 4-lb.	5 to 20-lb.	20 to 96-lb.	96 to 112-lb.
1 10	1 9	1 8

No. 52—5-gross, 11/- per gross.
No. 52—Combined Plain Bottle and Tie-over Jar, 12/- per gross.

No. 3—Hive, 10/6.

E. H. Taylor,
Wholesale and Retail Manufacturer,
Welwyn, Herts, and Johannesburg, S.A.

From: Beekeeping for Beginners
- according to the syllabus of the Board of Education for Schools
Walter Chitty F S Sc
1903

"Handle a book as a bee does a flower, extract its sweetness but do not damage it."
John Muir

DAY	JANUARY 2015 FORAGE	TEMP		WIND		CL'D	RAIN	1	2	3
		MIN	MAX	DIR	B.S			HIVE WEIGHT		
1										
2										
3										
4										
5										
6										
7										
8										
9										
10										
11										
12										
13										
14										
15										
16										
17										
18										
19										
20										
21										
22										
23										
24										
25										
26										
27										
28										
29										
30										
31										

JAN15

8,TH

1,TH
NEW YEAR'S DAY

9,FR

2,FR

10,SA
SR:08:03, SS:04:12

3,SA
SR:08:06, SS:04:03

11,SU

4,SU

12,MO

5,MO ○

13,TU

6,TU
EPIPHANY

14,WE

7,WE
ORTHODOX CHRISTMAS

15,TH

16,FR	24,SA SR:07:51, SS:04:34
17,SA SR:07:58, SS:04:22	25,SU BURN'S NIGHT
18,SU	**26,MO**
19,MO	**27,TU**
20,TU ✳	**28,WE**
21,WE	**29,TH**
22,TH	**30,FR**
23,FR	31,SA SR:07:41, SS:04:46

FEBRUARY

Winter either bites with its teeth or lashes with its tail.

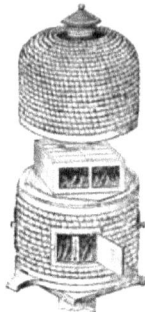
From: Bees and Bee-keeping, Vol 1. Scientific and Practical
- a complete treatise on the anatomy, physiology floral relations and
profitable management of the hive bee
by Frank R Cheshire FLS FRMS
Lecturer on Apiculture at South Kensington
1886

"In the village, a sage should go about
Like a bee, which, not harming
Flower, colour or scent,
Flies off with the nectar."
Anonymous, The Dhammapada

DAY	FEBRUARY 2015 FORAGE	TEMP		WIND		CL'D	RAIN	1	2	3
		MIN	MAX	DIR	B.S			HIVE WEIGHT		
1										
2										
3										
4										
5										
6										
7										
8										
9										
10										
11										
12										
13										
14										
15										
16										
17										
18										
19										
20										
21										
22										
23										
24										
25										
26										
27										
28										

FEB15

	8,SU
1,SU	9,MO
2,MO CANDLEMAS	10,TU
3,TU ○	11,WE
4,WE	12,TH
5,TH	13,FR
6,FR	14,SA SR:07:18, SS:05:12 ST VALENTINE'S DAY
7,SA SR:07:30, SS:04:59	15,SU

16,MO	**24,TU**
17,TU SHROVE TUESDAY	**25,WE**
18,WE ● ASH WEDNESDAY	**26,TH**
19,TH CHINESE NEW YEAR	**27,FR**
20,FR	28,SA SR:06:49, SS:05:37
21,SA SR:07:04, SS:05:25 SOMERSET CONFERENCE HOLDSWORTHY CONFERENCE	
22,SU	
23,MO	

MARCH

The louder the frog, the more the rain.

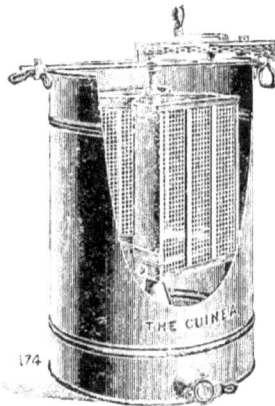

From: Beekeeping for Beginners
- according to the syllabus of the Board of Education for Schools
Walter Chitty F S Sc
1903

"The siren heralds a friend, the bee a stranger."
Hilda M Ransome, The Sacred Bee in Ancient Times and Folklore

DAY	MARCH 2015 FORAGE	TEMP MIN	MAX	WIND DIR	B.S	CL'D	RAIN	1	2	3 HIVE WEIGHT
1										
2										
3										
4										
5										
6										
7										
8										
9										
10										
11										
12										
13										
14										
15										
16										
17										
18										
19										
20										
21										
22										
23										
24										
25										
26										
27										
28										
29										
30										
31										

MAR15

	8,SU
1,SU ST DAVID'S DAY	**9,MO**
2,MO	**10,TU**
3,TU	**11,WE**
4,WE	**12,TH**
5,TH ○	**13,FR**
6,FR	**14,SA** SR:06:18, SS:06:01
7,SA SR:06:34, SS:05:49 BEE TRADEX, STONELEIGH	**15,SU** MOTHERING SUNDAY

16,MO	**24,TU**
17,TU ST PATRICK'S DAY	25,WE FEAST OF THE ANNUNCIATION
18,WE	**26,TH**
19,TH	**27,FR**
20,FR ◉ SPRING EQUINOX TOTAL SOLAR ECLIPSE GREENMOUNT CONFERENCE, NORTHERN IRELAND	28,SA SR:05:48, SS:06:25 NORTH OF ENGLAND CONFERENCE, NEWCASTLE
21,SA SR:06:02, SS:06:13 HINDU NEW YEAR	29,SU BRITISH SUMMERTIME COMMENCES PALM SUNDAY
22,SU	**30,MO** ST ALEXIUS DAY (UKRAINIAN BEEKEEPERS HANG ICONS OF THEIR PATRON SAINTS OF BEEKEEPING, ST SAVVATY AND ST ZOSIMA IN SHRINES AMONGST THEIR HIVES)
23,MO	**31,TU**

APRIL

If it thunders on All Fools' Day, it brings good crops of corn and hay.

Bees, Hives, AND ALL *Appliances.*

W. B. WEBSTER,

Expert (First Class) British Bee-Keepers' Association,

BINFIELD, BERKS.

SEND FOR ILLUSTRATED CATALOGUE, FREE BY POST

From: The Book of Beekeeping
by W B Webster
First Class Expert BBKA
1905

"My strength was renewed when I tasted a little honey."
1 Samuel 14:29

DAY	APRIL 2015 / FORAGE	TEMP		WIND		CL'D	RAIN	1	2	3
		MIN	MAX	DIR	B.S			HIVE WEIGHT		
1										
2										
3										
4										
5										
6										
7										
8										
9										
10										
11										
12										
13										
14										
15										
16										
17										
18										
19										
20										
21										
22										
23										
24										
25										
26										
27										
28										
29										
30										

APR15

	8,WE
1,WE	**9,TH**
2,TH MAUNDY THURSDAY	**10,FR**
3,FR **GOOD FRIDAY**	11,SA SR:06:15, SS:07:49
4,SA ○ SR:06:31, SS:07:37	12,SU EASTER (ORTHODOX)
5,SU EASTER SUNDAY	**13,MO**
6,MO BANK HOLIDAY (EASTER MONDAY - EXCL. SCOTLAND)	**14,TU** SIKH NEW YEAR
7,TU	**15,WE**

16,TH ●	**24,FR**
17,FR BBKA SPRING CONVENTION, HARPER ADAMS	25,SA SR:05:45, SS:08:12
18,SA ● SR:06:00, SS:08:00	26,SU
19,SU	**27,MO**
20,MO	**28,TU**
21,TU	**29,WE**
22,WE	**30,TH** ○
23,TH **ST. GEORGE'S DAY**	

MAY

A cow with its tail to the west, makes the weather best;
A cow with its tail to the east, makes weather the least.

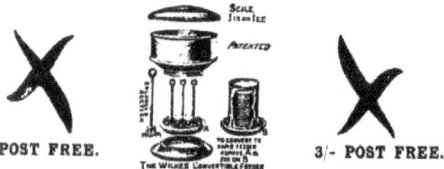

From: A Manual of Modern Beekeeping
IRISH BEE GUIDE
by J G Digges
1904

"Gracious words are a honeycomb,
sweet to the soul and healing to the bones."
Proverbs 16:24

DAY	MAY 2015 FORAGE	TEMP		WIND		CL'D	RAIN	1	2	3
		MIN	MAX	DIR	B.S			HIVE WEIGHT		
1										
2										
3										
4										
5										
6										
7										
8										
9										
10										
11										
12										
13										
14										
15										
16										
17										
18										
19										
20										
21										
22										
23										
24										
25										
26										
27										
28										
29										
30										
31										

MAY15

	8,FR
1,FR	**9,SA** SR:05:19, SS:08:35
2,SA SR:05:32, SS:08:23	10,SU ●
3,SU	**11,MO**
4,MO ○ BANK HOLIDAY	**12,TU**
5,TU	**13,WE**
6,WE	**14,TH** ASCENSION DAY
7,TH	**15,FR**

16,SA SR:05:08, SS:08:46	24,SU PENTECOST
17,SU	**25,MO** ○ SPRING BANK HOLIDAY
18,MO ◉	**26,TU**
19,TU	**27,WE**
20,WE	**28,TH**
21,TH	**29,FR**
22,FR	30,SA SR:04:51, SS:09:05
23,SA SR:04:59, SS:08:56	31,SU

JUNE

When windows won't open, and the salt clogs the shaker,
the weather will favour the umbrella maker!

From: Beekeeping Simplified for Cottager and Smallholder
by W Herrod-Hempsall
1915

"Tart words make no friends; a spoonful or honey will catch more flies
than a gallon of vinegar."
Benjamin Franklin

DAY	JUNE 2015 / FORAGE	TEMP		WIND		CL'D	RAIN	1	2	3
		MIN	MAX	DIR	B.S			HIVE WEIGHT		
1										
2										
3										
4										
5										
6										
7										
8										
9										
10										
11										
12										
13										
14										
15										
16										
17										
18										
19										
20										
21										
22										
23										
24										
25										
26										
27										
28										
29										
30										

JUN15

	8,MO
1,MO	**9,TU** ● ○
2,TU ○	**10,WE**
3,WE	**11,TH**
4,TH	**12,FR**
5,FR	13,SA SR:04:43, SS:09:18
6,SA SR:04:46, SS:09:12	14,SU
7,SU	**15,MO**

16,TU ◉	**24,WE** ○
17,WE	**25,TH**
18,TH RAMADAN	**26,FR**
19,FR	27,SA SR:04:45, SS:09:22
20,SA SR:04:43, SS:09:21	28,SU
21,SU SUMMER SOLSTICE FATHER'S DAY	**29,MO**
22,MO	**30,TU**
23,TU	

JULY

An early harvest is expected when the bramble blossoms early in June.

From: Beekeeping in War Time,
by W Herrod-Hempsall,
1919

"He is not worthy of the honey-comb,
That shuns the hives because the bees have stings."

DAY	JULY 2015 / FORAGE	TEMP		WIND		CL'D	RAIN	1	2	3
		MIN	MAX	DIR	B.S			HIVE WEIGHT		
1										
2										
3										
4										
5										
6										
7										
8										
9										
10										
11										
12										
13										
14										
15										
16										
17										
18										
19										
20										
21										
22										
23										
24										
25										
26										
27										
28										
29										
30										
31										

JUL15

	8,WE ●
1,WE	**9,TH**
2,TH ○	**10,FR**
3,FR	11,SA SR:04:46, SS:09:16
4,SA SR:04:49, SS:09:20	12,SU ORANGEMAN'S DAY
5,SU	**13,MO**
6,MO	**14,TU**
7,TU	**15,WE**

16,TH ☀	**24,FR**
17,FR	25,SA SR:05:13, SS:09:01
18,SA SR:05:04, SS:09:09	26,SU
19,SU	**27,MO**
20,MO	**28,TU**
21,TU	**29,WE**
22,WE	**30,TH**
23,TH RASTAFARI DAY	**31,FR**

AUGUST

No weather is ill if the weather is still.

From: Collected Leaflets on Beekeeping
Ministry of Agriculture and Fisheries
1922

"The bee is more honoured than other animals,
not because she labors, but because she labours for others."
St John Chrysostom - the Archbishop of Constantinople from 347 to 407

DAY	AUGUST 2015 FORAGE	TEMP MIN	MAX	WIND DIR	B.S	CL'D	RAIN	1	2	3 HIVE WEIGHT
1										
2										
3										
4										
5										
6										
7										
8										
9										
10										
11										
12										
13										
14										
15										
16										
17										
18										
19										
20										
21										
22										
23										
24										
25										
26										
27										
28										
29										
30										
31										

AUG15

	8,SA SR:05:34, SS:08:38
1,SA SR:05:23, SS:08:50	**9,SU**
2,SU	**10,MO**
3,MO SUMMER BANK HOLIDAY (SCOTLAND)	**11,TU**
4,TU	**12,WE** GROUSE SHOOTING BEGINS - MOVE BEES TO THE HEATHER PEAK OF PERSEIDS METEOR SHOWER (JULY 17TH - AUGUST 24TH)
5,WE	**13,TH**
6,TH	**14,FR** ●
7,FR	**15,SA** SR:05:45, SS:08:25 FEAST OF THE ASSUMPTION

16,SU	**24,MO** ST BARTHOLOMEW'S DAY (TRADITIONAL DAY FOR HARVESTING HONEY)
17,MO	**25,TU**
18,TU	**26,WE**
19,WE	**27,TH**
20,TH	**28,FR**
21,FR	29,SA ○ SR:06:07, SS:07:56
22,SA SR:05:56, SS:08:11	30,SU
23,SU	**31,MO** SUMMER BANK HOLIDAY

SEPTEMBER

When leaves fall early, autumn and winter will be mild;
when leave fall later, winter will be severe.

From: A Manual of Modern Beekeeping
IRISH BEE GUIDE
by J G Digges
1904

"Perfection in beekeeping is not found in a multiplicity of appliances, but in simplicity
and the elimination of everything not absolutely essential."
Brother Adam, In Search of the Best Strains of Bees

DAY	SEPTEMBER 2015 FORAGE	TEMP		WIND		CL'D	RAIN	1	2	3
		MIN	MAX	DIR	B.S			HIVE WEIGHT		
1										
2										
3										
4										
5										
6										
7										
8										
9										
10										
11										
12										
13										
14										
15										
16										
17										
18										
19										
20										
21										
22										
23										
24										
25										
26										
27										
28										
29										
30										

SEP15

	8,TU NATIVITY OF MARY
1,TU	**9,WE**
2,WE	**10,TH**
3,TH	**11,FR**
4,FR	12,SA SR:06:30, SS:07:25
5,SA SR:06:18, SS:07:40	13,SU ●
6,SU	**14,MO**
7,MO	**15,TU**

16,WE	**24,TH**
17,TH	**25,FR**
18,FR	26,SA SR:06:52, SS:06:52
19,SA SR:06:41, SS:07:08	27,SU
20,SU	**28,MO** ○
21,MO	**29, TU**
22,TU	**30, WE**
23,WE YOM KIPPUR EQUINOX	

OCTOBER

Thunder in the Fall foretells a cold winter.

From: Practical Beekeeping
- being plain instructions to the amateur for the successful management
of the honey bee
by Frank Cheshire
Editor of the Apiary Department of "The Country"

"Until you have smoked out the bees, you can't eat the honey."
Russian Proverb

DAY	OCTOBER 2015 FORAGE	TEMP MIN	TEMP MAX	WIND DIR	WIND B.S	CL'D	RAIN	1	2	3
								HIVE WEIGHT		
1										
2										
3										
4										
5										
6										
7										
8										
9										
10										
11										
12										
13										
14										
15										
16										
17										
18										
19										
20										
21										
22										
23										
24										
25										
26										
27										
28										
29										
30										
31										

OCT15

8,TH	
1,TH	**9,FR**
2,FR	10,SA SR:07:15, SS:06:21
3,SA SR:07:03, SS:06:36	11,SU
4,SU ST FRANCIS ANIMAL DAY	**12,MO**
5,MO	**13,TU** ●
6,TU	**14,WE**
7,WE	**15,TH**

16,FR	24,SA SR:07:39, SS:05:51
17,SA SR:07:27, SS:06:05	25,SU DAYLIGHT SAVING TIME ENDS
18,SU	**26,MO**
19,MO	**27,TU** ○
20,TU	**28,WE**
21,WE	**29,TH** **NATIONAL HONEY SHOW, WEYBRIIDGE**
22,TH	**30,FR**
23,FR	31,SA SR:06:51, SS:04:37 ALL HALLOW'S EVE

NOVEMBER

Sounds travelling far and wide,
A rainy day will betide.

W. HOLLANDS,

Manufacturer of Movable Comb Hives,

And all other Appliances for the Easy and Successful
Management of the Honey Bee.

HAVING been favoured with Patterns and Instructions, I intend to make the CHESHIRE HIVE, and other Inventions of MR. CHESHIRE, a Specialité. All the latest Improvements of any value have been added to these Hives; they are so constructed as to be suitable for either simple or the most advanced Bee-keeping. Made in the best manner, of well-seasoned materials, and having a Zinc Roof, they are well nigh everlasting.

Every Description of APICULTURAL NECESSARIES always on Hand.

STOCKS AND SWARMS OF BEES; Prices according to Season.

Head Quarters for the CHESHIRE CURE FOR FOUL BROOD. Wholesale and Retail Dealers supplied; Prices on application. Per Single Bottle, with Instructions, post free, 1s. 3d.

Agent for the Best FRENCH WHITE GLASS BOTTLES, either with or without Screw Capsules.

SEND TWO STAMPS FOR CATALOGUE.

Note the Address:

W. HOLLANDS, THE APIARY, WADDON, CROYDON.

TERMS—Strictly Cash with Order, or on Receipt of Invoice.

C. T. OVERTON,

EXPERT OF THE SUSSEX BEE-KEEPERS' ASSOCIATION,

Offers to Bee-keepers APIARIAN SUPPLIES of unsurpassed Quality and fine Workmanship, consisting of

Hives, Supers, Comb Foundation, Sections, Feeders, Bee Veils, Smokers and Extractors, &c., &c.

PRIME SWARMS OF LIGURIAN AND BLACK BEES.
ORDERS RECEIVED FOR PURE IMPORTED LIGURIAN QUEENS.

Send Two Stamps for Illustrated Catalogue to

CHARLES T. OVERTON, "LOWFIELD APIARY," CRAWLEY, SUSSEX.

CARNIOLAN, CYPRIAN, and LIGURIAN QUEENS.

The best obtainable, either Home-bred or Imported.

HOME-BRED QUEENS of either race, raised from selected mothers, at an Apiary where no other Bees exist within a radius of several miles; hence a large percentage of these Queens prove PURELY FERTILISED.

Maker of CHEAP PRACTICAL HIVES and other Apicultural Appliances.
AWARDED PRIZES at every Show where Exhibited last season.

"MODERN ❖ BEE-KEEPING ❖ APPLIANCES."
Post Free, 1½d., contains useful Information for all Bee-keepers.

Address:

J. R. W. HOLE, TARRINGTON, LEDBURY.

From: Bees and Bee-keeping, Vol 1. Scientific and Practical
- a complete treatise on the anatomy, physiology floral relations and profitable
management of the hive bee
by Frank R Cheshire FLS FRMS
Lecturer on Apiculture at South Kensington
1886

"There are certain pursuits which, if not wholly poetic and true, do at least suggest a
nobler and finer relation to nature than we know. The keeping of bees, for instance."
Henry David Thoreau

DAY	NOVEMBER 2015 FORAGE	TEMP MIN	MAX	WIND DIR	B.S	CL'D	RAIN	1	2	3 HIVE WEIGHT
1										
2										
3										
4										
5										
6										
7										
8										
9										
10										
11										
12										
13										
14										
15										
16										
17										
18										
19										
20										
21										
22										
23										
24										
25										
26										
27										
28										
29										
30										

NOV15

	8,SU REMEMBRANCE SUNDAY
1,SU ALL SAINTS' DAY	**9,MO**
2,MO ALL SOULS' DAY	**10,TU**
3,TU	**11,WE** ●
4,WE	**12,TH**
5,TH GUY FAWKES DAY	**13,FR**
6,FR ●	**14,SA** SR:07:16, SS:04:05
7,SA SR:07:03, SS:04:25	**15,SU**

16,MO	24,TU
17,TU	25,WE THANKSGIVING DAY (USA)
18,WE	26,TH
19,TH	27,FR
20,FR	28,SA SR:07:38, SS:03:58
21,SA SR:07:27, SS:04:05	29,SU ADVENT BEGINS
22,SU	30,MO ST ANDREW'S DAY
23,MO	

DECEMBER

Sharp horns on the moon threaten bad weather.

From: Beekeeping Simplified for Cottager and Smallholder
by W Herrod-Hempsall
1915

"We ought to do good to others as simply as a horse runs,
or a bee makes honey, or a vine bears grapes season after season
without thinking of the grapes it has borne."
Marcus Aurelius

DAY	DECEMBER 2015 / FORAGE	TEMP MIN	MAX	WIND DIR	B.S	CL'D	RAIN	1	2	3
								HIVE WEIGHT		
1										
2										
3										
4										
5										
6										
7										
8										
9										
10										
11										
12										
13										
14										
15										
16										
17										
18										
19										
20										
21										
22										
23										
24										
25										
26										
27										
28										
29										
30										
31										

DEC15

	8,TU
1,TU	**9,WE**
2,WE	**10,TH**
3,TH	**11,FR** ●
4,FR	12,SA SR:07:56, SS:03:51
5,SA SR:07:48, SS:03:53	13,SU
6,SU ST NICHOLAS' DAY FIRST DAY OF HANNUKAH	**14,MO** LAST DAY OF HANNUKAH
7,MO ST AMBROSE DAY (PATRON SAINT OF BEEKEEPERS)	**15,TU**

16,WE	24,TH
17,TH	25,FR ○ CHRISTMAS DAY
18,FR	26,SA SR:08:05, SS:03:56 BOXING DAY
19,SA SR:08:02, SS:03:52	27,SU
20,SU	28,MO
21,MO WINTER SOLSTICE	29, TU
22,TU	30, WE
23,WE	31,TH NEW YEAR'S EVE

Hive/ Q NO.	Year Q Raised	Frames of Brood Autumn 2014	Combs Covered	Honey Stored- Sugar fed Kg	Combs Covered Spring 2015	Frames of Brood Spring 2015	Spring Feeding Kg	Queens Reared	Nuclei
1									
2									
3									
4									
5									
6									
7									
8									
9									
10									
11									
12									
13									
14									
15									
16									
17									
18									
19									
20									
21									
22									
23									
24									

HONEYBEE COLONIES

1									
2									
3									
4									
5									
6									
7									
8									
9									
10									
11									
12									
13									
14									
15									
16									
17									
18									
19									
20									
21									
22									
23									
24									

BEEEKEEPING RECORDS

Number	items	Est. Value £	P
	Stocks of Bees		
	Empty Hives		
	Combs - Deep - Shallow		
	Frames		
	Foundations		
	Honey Extractor		
	Honey Tanks		
	Other items		
	Honey Jars		
	Honey		

JANUARY 2016							FEBRUARY 2016							MARCH 2016					
S	M	T	W	T	F	S	S	M	T	W	T	F	S	S	M	T	W	T	F	S
					1	2		1	2	3	4	5	6			1	2	3	4	5
3	4	5	6	7	8	9	7	8	9	10	11	12	13	6	7	8	9	10	11	12
10	11	12	13	14	15	16	14	15	16	17	18	19	20	13	14	15	16	17	18	19
17	18	19	20	21	22	23	21	22	23	24	25	26	27	20	21	22	23	24	25	26
24	25	26	27	28	29	30	28	29						27	28	29	30	31		
31																				

APRIL 2016							MAY 2016							JUNE 2016					
S	M	T	W	T	F	S	S	M	T	W	T	F	S	S	M	T	W	T	F	S
					1	2	1	2	3	4	5	6	7				1	2	3	4
3	4	5	6	7	8	9	8	9	10	11	12	13	14	6	7	8	9	10	11	
10	11	12	13	14	15	16	15	16	17	18	19	20	21	13	14	15	16	17	18	
17	18	19	20	21	22	23	22	23	24	25	26	27	28	20	21	22	23	24	25	
24	25	26	27	28	29	30	29	30	31					27	28	29	30			

JULY 2016							AUGUST 2016							SEPTEMBER 2016					
S	M	T	W	T	F	S	S	M	T	W	T	F	S	S	M	T	W	T	F	S
					1	2		1	2	3	4	5	6					1	2	3
3	4	5	6	7	8	9	7	8	9	10	11	12	13	4	5	6	7	8	9	10
10	11	12	13	14	15	16	14	15	16	17	18	19	20	11	12	13	14	15	16	17
17	18	19	20	21	22	23	21	22	23	24	25	26	27	18	19	20	21	22	23	24
24	25	26	27	28	29	30	28	29	30	31				25	26	27	28	29	30	
31																				

OCTOBER 2016							NOVEMBER 2016							DECEMBER 2016					
S	M	T	W	T	F	S	S	M	T	W	T	F	S	S	M	T	W	T	F	S
						1		1	2	3	4	5						1	2	3
2	3	4	5	6	7	8	6	7	8	9	10	11	12	4	5	6	7	8	9	10
9	10	11	12	13	14	15	13	14	15	16	17	18	19	11	12	13	14	15	16	17
16	17	18	19	20	21	22	20	21	22	23	24	25	26	18	19	20	21	22	23	24
23	24	25	26	27	28	29	27	28	29	30				25	26	27	28	29	30	31
30	31																			

THE BRITISH BEEKEEPERS ASSOCIATION · FOUNDED 1874 ·

Spring 2015 Convention

Lectures
Courses
Workshops
Trade-show
Accommodation
Dinners

Tickets on sale and
on-line bookings
January 2015

Harper Adams University

Newport, Shropshire TF10 8NB

Friday 17 April • Saturday 18th April • Sunday 19th April 2015

enquiries to: tim.lovett@bbka.org.uk

Every effort is made to keep entries up to date but the publishers cannot be held responsible for errors or omissions.
Associations and all other groups listed have been requested (August 2014) to supply updated entries.
Readers who are aware of inaccuracies are asked to send updates to jerry@northernbeebooks.co.uk

DIRECTORY, Associations and Services

Ninemaidens

Mead

Award winning mead & honey

visit **www.ninemaidensmead.com**
or tel. 01209 820939 / 860630

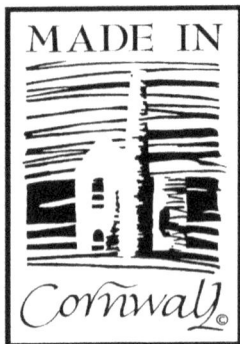

MADE IN

APPROVED ORIGIN SCHEME

CORNWALL COUNCIL

Cornwall©

DIRECTORY, ASSOCIATIONS AND SERVICES

BEE-EDU

✉ ☎

Bee Educated (BEE-EDU)
e-Learning for Beekeepers

-†beekeeping virtual classroom.

Contact Us
Bee-Edu Administrator
Steven Turner
Email: st@zbee.com

Bee-Edu is a Moodle website set-up specifically for beekeeping tutors and their students.

What is Moodle?

Moodle is a Virtual Learning Environment (VLE) which makes it easy for tutors to provide online support for†their course. It provides†a central space on the web where students can access a set of tools and resources at all times.

What are the Benefits?

1. It is an easy way to communicate with students: The course news automatically emails messages to all students. Forums can also be used to answer commonly asked questions, to provide a space for informal peer to peer student discussion or even online tutorials.

2. It is at quick way to share documents: Moodle provides a place where you can easily create web pages with information about your course and provide links to word, PDF documents, slides, and other resources that your students will want to access.

3. It has easy access to relevant and useful online resources: There is so much information about beekeeping on the Internet which makes it difficult for students to find reliable and trusted resources. You can use your Moodle to provide links directly to these resources in an organised way.

4. Online assignment handling: Online assignment handling can save time and effort for everyone involved, whether it†s just used for student submission with marking

done on paper or the whole process is moved online saving time, postage and paper.

5. Other advantages: Making resources available online can save time and money in photocopying. For example keep a central copy of documents†online so everyone can find the latest version of a course handbook etc. Provide handouts online so that students only print out what they really need. It is easy to experiment with new ideas and tools, it's a low risk way to incorporate new tools and ideas into your teaching. Tutors can manage their materials. If all your course information is on Moodle this is easy access this year on year.

6. Other features and tools: Course calendar: use this to flag important events to everyone on your course. Profiles and contact information helps students and tutors get to know each other from the start of the course. Deliver content: add slides and photographs. Video and audio: many tutors find it easy to record lectures as podcasts or even arrange for videos of lectures, posting these online and making it available to students is straight forward with Moodle.

7. Group tools for students. There are many tools that students can use for collaboration with each other such as forums, wiki and chat.

Beekeeping Course Tutors
Bee-Edu are providing a free virtual learning environment (VLE) for beekeeping tutors to complement their existing courses or build new online courses in beekeeping related topics.

Bee-Edu will support those individuals or associations that are keen to try this medium but help will be limited to testing and guidance with the technology rather than organising the teaching resources. There is no time scale and I will not be imposing any rules on tutors who wish do their own thing.

Students
It is†expected that most courses created on Bee-Edu would need enrolment keys and their students would be directed to the site by their tutors.

BDI

BEE DISEASES INSURANCE LTD

B D I

SECRETARY
Donald Robertson-Adams
Ffosyffin, Ffostrasol,
Llandysul, Ceredigion,
SA44 5JY
07532 336076
secretary@beediseases-
insurance.co.uk

TREASURER AND
SCHEME B MANAGER
Mrs Sharon Blake
Stratton Court,
South Petherton,
Somerset TA13 5LQ
01460 242124
treasurer@beediseases-
insurance.co.uk

CLAIMS MANAGER
Bernard Diaper
57 Marfield Close,
Walmley,
Sutton Coldfield B76 1YD
07711456932
claims@beediseasesin-
surance.co.uk

PRESIDENT
Martin Smith
137 Blaguegate Lane
Lathom, Sklemersdale
Lancs WN8 8TX
07831 695732
president@beediseases-
insurance.co.uk

BDI is a small insurance company that specialises in compensating beekeepers in England and Wales, who have had their by the Bee Inspector as a result of being infected by a notifiable disease. These currently are European Foul Brood (EFB) and American Foul Brood (AFB). as the result of being infested by a notifiable pest (Small Hive Beetle and Tropilaelaps), should they reach England or Wales.

The company is regulated as an insurance company by the Prudential Regulatory Authority and supervised by the Prudential Regulatory Authority and the Financial Conduct Authority.

BDI is owned by a number of BKAs who are their members. There are no full time employees or premises. BDI is run by a small group of officers on a day – to – day basis from their homes. In addition there is a board of directors who meet regularly.

BDI subscriptions are paid along with the local BKA subscription. This is compulsory if your BKA is a BDI member. You will also be asked to pay premiums for the number of additional colonies you expect to have during the year, above the basic free colonies although you can also top-up during a year.

Further details of current subscription and premium rates together with compensation rates are available on the BDI website.

Research Grants

BDI supports research into the causes of honey bee diseases in a number of ways including the industry

116

sponsorship of PhD studentships. Details of the current and past research activities that have been supported by BDI are available on our website.

Training Grants
During 2014, BDI offered a grant of £100 to local associations who organise a practical event, with leaders/demonstrators who are knowledgeable in disease recognition. BDI were pleased to be working with the National Bee Unit to promote bee health. These events covered a range of topics, including for example, maintaining healthy bees, disease recognition, integrated pest management and approved medicines. It is the plan to repeat this grant in the 2015 season.

Further details of the full range of activities carried out can be found on the BDI website HYPERLINK "http://www.beediseasesinsurance. co.uk/"www.beediseasesinsurance.co.uk, or contact via secretary@ beediseasesinsurance.co.uk

Bee Farmers Assocation Limited

BEE FARMERS ASSOCIATION

The BFA represents the professional beekeepers of the UK.
www.beefarmers.co.uk

CHAIRMAN
Murray McGregor
Denrosa Apiaries
Victoria Street
Coupar Angus
Perthshire
Scotland
PH13 9AE
01828 627721
chair@beefarmers.co.uk

VICE-CHAIR
David Wainwright
Tropical Forest Products
Box 92
Aberystwyth
Dyfed
Wales
SY23 1AA
01970 832511
deputychair@beefarmers.
co.uk

COMPANY SECRETARY
John Howat
8 Olivers Close
West Totton
Southampton
Hampshire
SO40 8FH
023 8090 7850
honsec@beefarmers.co.uk

FINANCE DIRECTOR
John Heard
36 The Green
Long Whatton
Loughborough
Leicestershire
LE12 5DB
01509 646767
financedirector@
beefarmers.co.uk

As the professional trade association for the sector, we represent around 400 bee farming businesses. Our members produce honey throughout Great Britain and supply products bulk, wholesale and retail. The association is the largest contract pollinator in the UK.

A significant number of members are employed as bee inspectors, responsible for identifying and dealing with notifiable disease.

We have one business meeting a year which follows the Annual General Meeting, held in the spring to coincide with one of the major trade events. There are also twice-yearly regional meetings, usually featuring guest speakers.

The Bee Farmers' Association works with the National Farmers' Union (NFU) and the Honey (Packers) Association to promote ecologically sensitive farming and consumer awareness.

The work of the Bee Farmers' Association

- To monitor and to keep members informed about developments in commercial beekeeping, bee science and UK and EU legislation.
- Liaison with farmers, growers, contractors, consumers and other organisations.
- Liaison and cooperation with UK beekeeping organisations.
- Liaison with UK Government departments dealing with beekeeping, medicines and allied matters.
- Contact with European beekeeping organisations and representation on the EU Honey Working Party (COPA/COGECA) in Brussels.
- Member of the EU Honey Task Force.
- Political lobbying through MPs and MEPs.
- Member of the Confederation of National Beekeeping Associations (CONBA).
- Associate member of the Honey Association.
- Member of the National Farmers' Union.

Benefits of membership

- Advice and support on all aspects of honey farming and commercial beekeeping.
- Networking opportunities with others in the sector.
- Insurance for products, third-party liability, and employer's liability.
- Association journal, featuring informative articles, case studies, news and updates on meetings with Defra, Fera, VMD and the EU, reports on current beekeeping issues and commercial developments worldwide.
- Free classified advertising and discounted display advertising in the association journal.
- Sources of equipment and sundries; product directory of specialist suppliers.
- Bulk purchase scheme and supplier discounts.
- Regional meetings which provide networking and trading opportunities.
- Spring Convention for members and partners, held each spring in the UK or abroad; includes visits to bee farms and research establishments; lectures and discussions on bee-related matters; sight-seeing and social events.
- Eligibility for the Disease Accreditation Scheme for Honeybees (DASH).
- Free circulation of UK and foreign beekeeping journals.
- Crop and winter loss reports.
- Pollination contracts.

For Membership

Members are expected to have a good level of competence as beekeepers and to maintain a professional approach in their operations.

Full membership is available to those with 40 or more production colonies.

Associate membership is available to those with 20-39 production colonies who wish to progress to operation on a semi-commercial or commercial scale.

Corporate membership

Please contact the General Secretary for details.

How to apply

Application forms are available to download from the Bee Farmers' Association website, or from the Membership and Administration Officer.

GENERAL SECRETARY
Margaret Ginman MBA FRSA
Hendal House
Groombridge
Kent
TN3 9NT
01892 864499, 07795 153765
gensec@beefarmers.co.uk

Enquiries in relation to: European Union (EU) and government representation, strategic partnerships, corporate sponsorship, press and public relations, apprenticeship scheme.

MEMBERSHIP AND ADMINISTRATION
Alex Ellis
23 Edgeley Road
Whitchurch
Shropshire
SY13 1EU
01948 510726, 07972 281496
admin@beefarmers.co.uk

Enquiries in relation to: membership, communications, publications, insurance, events and training, Disease Accreditation Scheme for Honeybees (DASH).

POLLINATION SECRETARY
Alan Hart
61 Fakenham Road
Great Witchingham
Norwich
Norfolk
NR9 5AE
01603 308911
pollination@beefarmers.co.uk

GIANT
BEEKEEPING SALE

Only £5 per adult, family tickets available, children go free

Lectures from **DEFRA** on disease management, etc

Over thirty traders including NBB, Thornes, Maisemore, Paynes, Swienty, BBWear, Sherriff, ModernBeekeeping,etc

**Come and grab an early bargain,
see what's new in the world of beekeeping,**

Saturday 7th March 2015
9am to 4:30pm

Hall H1
Stoneleigh Park
Warwickshire
CV8 2LG

www.beetradex.co.uk
info@beetradex.co.uk

Or write
Beetradex Ltd
Unit 5 Maesbury Road Industrial Estate
Oswestry
Shropshire
SY10 8HA

BEEKEEPING EDITORS' EXCHANGE SCHEME

BEES is a self-help grouping of local, county and country beekeeping association editors, which operates principally by exchanging journals through a central address. The scheme is supported by Northern Bee Books.

BEES was founded in 1984 and for many years has been an exchange of paper copy. However, the focus has now changed to an electronic exchange, using the server of one of the participating editors.

Now fully established as part of the British and Irish beekeeping scene, the scheme brings up to date information to beekeepers throughout the British Isles.

B.E.E.S
Helping Editors
Help Themselves

Sponsored by
NORTHERN BEE BOOKS

The aims are:
- to exchange ideas for content and production methods
- to aid others by experience
- to communicate matters editorial
- to share information on national beekeeping issues
- to help and reassure those new to the task
- to give a wider readership to the best writing in beekeeping journalism

If you are an editor or potential editor and would like to know more about how we operate write to Chris Jackson
22 Chapter Close, Oakwood, Derby, DE21 2BG
editors-owner@ebees.org.uk

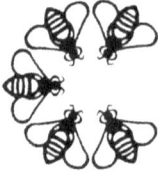

BEES ABROAD UK Ltd

Relieving Poverty through Beekeeping

ADMINISTRATOR:
MRS VERONICA BROWN,
PO Box 2058,
Thornbury
Bristol
BS35 9AF.
0117 230 0231
info@beesabroad.org.uk

Bees Abroad is a UK-registered charity (No 1108464) established in 1999. Its principle aim is the relief of poverty in the developing world using beekeeping and associated skills as tools of individual, group and community empowerment for poverty alleviation and to provide sustainable income. Beekeeping is a valuable tool as it is socially and culturally acceptable for both genders across a wide age range. It can cost very little to set up a beekeeping operation, which will deliver benefits for income, education, health, environment and community. Beekeeping and its associated skills deliver access to gainful self-employment for poor and disadvantaged groups. This enables them to recover social status, improve social interactions, obtain income and acquire new skills to build the confidence to represent their own interests. Bees Abroad receives a high volume of direct appeals for assistance from groups all over the world. In practice, it achieves its aims through a volunteer network of supporters, committee members and project managers. Bees Abroad takes care to ensure that its projects are sustainable and not dependent on constant external input. This is done by supporting community group initiatives, setting up village-based field extension services, running training courses for beekeeping trainers and financing local trainers' wages. All Bees Abroad projects are designed to become self-financing after a defined time period. Its first two projects in Nepal and Cameroon now employ 42 beekeeper trainers and involve many more beekeepers. It currently has projects either running or seeking funding in Cameroon, Ghana, Kenya, Liberia, Malawi, Nigeria, Uganda, Yemen and Zambia.

Bees Abroad is run by volunteers who are all beekeepers. They currently undertake all activities, including fundraising, though an administrator is employed for one day a week. It also arranges beekeeping holidays to a variety of locations, including Morocco, Poland, Chile and Nepal.

Bees Abroad is delighted to have the support of its patrons: The Most Reverend Justin Welby, Archbishop of Canterbury, Jimmy Doherty (Jimmy's Farm), Martha Kearny (Broadcaster and Journalist), Adam Hart (Professor of Science Communication, University of Gloucestershire), Michael Badger, MBE (Past President, BBKA), Brian Sherriff (BJ Sherriff International), Bill and Paula Stevens (National Bee Supplies), Eric Hiam (Maisemore Apiaries) and Richard Jones (Past Director, IBRA).

For more details of what we do and how you can help, contact Veronica Brown, the Administrator, Bees Abroad (info@beesabroad.org.uk). You can learn more about our work and make a regular or one-off donation through our website, www.beesabroad.org.uk

BEES *for* DEVELOPMENT TRUST

supporting beekeepers in developing countries
www.beesfordevelopment.org

Bees *for* Development
1 Agincourt Street, Monmouth, NP25 3DZ, UK
Tel +44 (0)16007 14848
info@beesfordevelopment.org

YOU CAN HELP US BY:

- **Ensuring** that your group or organisation knows about our work, and supports us if possible
- **Making** a gift of a Resource Box for a training course in a school or project
- **Subscribing** to our magazine, BfD Journal, or Sponsoring a subscription to the Journal for a beekeeper working in a poor country
- **Helping** us to represent our organisation at events
- **Offering** your skills to work with us as a volunteer
- **Giving** a donation
- **Joining** one of our Beekeepers' Safaris
- **Using** our tamper proof seals or labels when you sell or gift your honey
- **Buying** from our special shop in Monmouth or from our on-line store. Proceeds from sales go to support our charitable work.
- **Attending** one of our unique training Courses in UK on Sustainable Beekeeping or on *Strengthening Livelihoods by Means of Beekeeping*. We also run courses at *Humble by Nature* and at *Ragman's Permaculture Farm* in the Wye Valley.

What we do

Bees *for* Development encourages beekeeping to alleviate poverty and to maintain biodiversity in developing countries.

We run practical, community-based projects to develop the skills and knowledge that enable families in poor areas to create reliable income from bees.

We provide educational resources and promote awareness of biodiversity.

Bees *for* Development Journal provides readers in over 130 countries with practical advice, information and articles about how they can keep bees sustainably and increase their incomes.

We always need more help and skills please contact us if you might like to become involved. We are a highly professional organisation, working in the beekeeping development sector for more than 21 years. We are respected and trusted by beekeepers world-wide. Please encourage your beekeeping group or other organisation to support our work: do contact us for more details of our work and current projects.

Charity 1078803

BRITISH BEEKEEPERS' ASSOCIATION

www.bbka.org.uk

COMMITTEES OF THE EXECUTIVE AND SECRETARIES

Education & Training

The development of information from practical guidance notes, advisory leaflets, training materials while also undertaking it's own educational initiatives in support of improving the knowledge and skills of beekeepers at all levels. Education & Training liase with the Examination Board to develop training materials to support Association tutors with products such as the Course in a Case.

Examination Board

The BBKA examination board provide a structured range of examinations fulfilling the needs of all beekeepers from Junior Certificate to Master Beekeeper. The board are responsible for all matters relating to the syllabus, content and assessment and operate independently of the BBKA board of Trustees. Where Associations have no Examinations Secretary the Association Secretary deals with examinations. To help future candidates it is suggested that Associations without an Examination Secretary appoint one. Associations are responsible for arranging a suitable room for the written examinations and recommending an invigilator.
Contact Val Frances, Exam Board Secretary
Email: val.frances@bbka.org.uk Tel 01226 286341

FINANCE

This team of Trustees reviews & agrees all budgets, handles all investment matters, finalising insurance policies and sets proposals relating to capitation.

OPERATIONS DIRECTOR &
GENERAL SECRETARY
Jane Moseley
The British Beekeepers'
Association,
National Beekeeping Centre,
Stoneleigh Park
Warwickshire CV8 2LG
024 7669 6679
Fax: 024 7669 0682

BRITISH BEEKEEPERS
ASSOCIATION
National Beekeeping Centre
Stoneleigh Park,
Kenilworth, Warks CV8 2LG
02476 696679
Fax: 024 7669 0682
Office hours 9.00am–5.00 pm
Monday - Friday (inclusive)

BBKA

Governance

Primary areas of responsibility are to ensure that we adhere to Charity Commission rules, that we operate within the constitution in addition to ensuring that our Trustees act in the best interests of the BBKA and it's members.

Operations & Membership Services

This team ensures that all Membership Services are administered effectively and on time and that the organisation operates efficiently. It also acts as a co-ordinator for all external fundraising.
Contact: jane.moseley@bbka.org.uk

Public Affairs

Whether it be government liason, both UK & EU, or press activity this comes from the Public Affairs team. All enquiries should be made to BBKA Press Officer: gill.maclean@bbka.org.uk

Technical & Environmental

All technical issues and their their potential impact on bees and beekeeping are assessed and monitored within this team. All research projects are reviewed and recommendations made by Technical & Environmental group.

Insurance

Members of BBKA, Area Associations and officials are indemnified against claims for Public Liability to a limit of £10million, Product Liability to a limit of £10 million, Professional Indemnity to a limit of £2 million relating to their beekeeping activities. BBKA Association Officer and Trustee liability insurance also applies to a limit of £10 million. Each new claim carries an excess payable by the member.

An 'All Risks' policy is available to both individuals and Associations, to cover the loss or damage of property & equipment. Details are available via www.bbka.org.uk or the main office: 02476 696679

Publications

- BBKA News is issued monthly free to all members of the BBKA, featuring articles about bees, beekeeping and the other associated articles of interest. Editorial: editorial@bbkanew.org.uk Advertising: advertising@bbkanews.org.uk

- BBKA Year Book is published each Spring and is for Association use and reference.It contains detailed information on the BBKA including useful reference tools such as a directory of Lecturers and Demonstrators.

- Members Handbook is published annually and sent to new members

- BBKA Introduction to Beekeeping

BBKA Website - www.bbka.org.uk

The BBKA Website contains technical information, is easy to navigate and supports both beekeepers and the general public. You can download publications, find help and advice in the discussion forums, purchase merchandise, learn about Bees, use the Bees4kids section, download BBKA exam application forms and the exam syllabus. Within the Members Only area, specific insurance downloads and other member only information is available. Associations beekeeping events are promoted.

A Swarm Collector database is included within the site enabling the general public with a direct link to a local swarm collector.

Events

Area and local associations attend and exhibit at various events within their local throughout the year while the BBKA supports selected national shows. Whether it be village fete or national exhibition these events continue to provide a vital service for the dissemination of knowledge.

BBKA

BBKA Spring Convention
Held in April every year this is a firmly established major beekeeping event. Lectures and Workshops are staged over 3 days with a trade exhibition. Both Friday and Sunday are member only days which are ticketed.

Slide Library
The BBKA slide library has been digitised for ease of use and preservation. For a list of slides available and their format please go the BBKA Members Area at www.bbka.org.uk or contact the BBKA office.

Subscriptions & Membership Fees
Individual Membership of the BBKA is £38 per annum, for an Overseas Member the fee is £28.00. All other membership is via local associations.
Friends Membership is also available via www.bbka.org.uk

Exam Board Footnote
Where Associations have no Examinations Secretary the Association Secretary deals with examinations. To help future candidates it is suggested that Associations without an Examination Secretary appoint one. Associations are responsible for arranging a suitable room for the written examinations and recommending an invigilator.

If you live in an area without a nominated Exam Secretary, you should contact Val Frances, Exam Board Secretary Email: val.frances@bbka.org.uk Tel 01226 286341

BBKA Enterprises
BBKA Enterprises Ltd is a private company, limited by guarantee with all profits from the trading activities being donated to the BBKA. Via the BBKA online shop a range of beekeeping, corporate and related items, specially selected books, gifts, travel items and educational materials are available.

Visit www.thepollenbasket.com, the official BBKA web shop, or call 02476 696679

SOME ASSOCIATION SECRETARIES

Cambridgeshire Beekeepers
Susanna Fenwick
susfen@hotmail.com
http://www.cbka.org.uk

Essex
Michael Webb
secretary@ebka.org
http://ebka.org

Hertfordshire
John Palombo
secretary@hertsbees.org.uk
http://hertsbees.org.uk

Huntingdonshire
Nicholas Steiger
secretary@huntsbka.org.uk
http://www.huntsbka.org.uk

Norfolk
Louise Hutchinson
contact@norfolk
beekeepers.co.uk
http://www.norfolk
beekeepers.co.uk

**Norfolk
West & Kings**
Lynn Kay Marshall
sec@wnklba.co.uk http://
www.wnklba.co.uk

Peterborough & District
George Newton
george_newton1@hotmail.
com
http://www.bbka.org.uk/
local/peterborough/

Suffolk
Ian McQueen
jackie.mcqueen@ntlworld.
com
http://www.suffolkbeekeep-
ers.co.uk

Chesterfield & District
Robin Bagnall
robin.bagnall@virgin.net
http://www.cdbka.org.uk

Derbyshire
Mike Cross
crosssk@btinternet.com
http://www.derbyshire-bka.
org.uk

Lincolnshire
Celia Smith
sec.lincsbka@yahoo.com
http://www.bbka.org.uk/
local/lincolnshire/

Northamptonshire
Ruth Stewart
secretary@northantsbees.
org.uk
http://www.northantsbees.
org.uk

Nottinghamshire
Maurice Jordan
mauricejordan11@btinternet.
com
http://www.bbka.org.uk/
local/nottsbees/

Rutland
Fliss Haynes
fliss@rbka.org
http://www.rbka.org

Gwent
William Stewart
secretary@gwentbeekeep-
ers.co.uk
http://www.gwentbeekeep-
ers.co.uk

Isle of Man
Janet Thompson
theman@iombeekeepers.
com
http://www.iombeekeepers.
com

Isle of Wight
Frank Stevens
frankndot29@ymail.com
http://www.iwbka.org.uk

London
Angela Woods
sec@lbka.org.uk
http://www.lbka.org.uk/
index.php

Middlesex
Mary Hunter
mary@hunter67.myzen.co.uk
http://www.prbka.co.uk

Twickenham & Thames Valley
Sara Crofton
sarah.crofton@btconnect.
com
http://www.twickenham-
bees.org.uk

BBKA

Cleveland
Steve Jacklin
steve@jacklin.cx
http://www.teesbees.co.uk

Durham
Lynn Ramsay
lynnramsay@hotmail.co.uk
http://www.bbka.org.uk/
local/durham/index.shtml

Harrogate & Ripon
Janet Paley
secretary@hrbka.org.uk
http://www.hrbka.org.uk

Newcastle & District
Brian Diver
bdiver@accesstraining.org
http://www.bbka.org.uk/
local/newcastle/

Northumberland
Revd Benjamin Hopkinson
benjamin.hopkinson@gmail.
com
http://www.bbka.org.uk/lo-
cal/northumberland/branch-
es/hexham/index.shtml

Cheshire Beekeepers
Elizabeth Camm
thesecretary@cheshire-bka.
co.uk
http://www.cheshire-bka.
co.uk

Kendal & South Westmorland
Peter Llewellyn
pdwllewellyn@yahoo.co.uk
http://www.kendalbeekeep-
ers.org.uk

Lancashire & North West
Martin Smith
ormskirk_beekeepers@
hotmail.com
http://www.lancashirebee-
keepers.org.uk

Lancaster
Pip Merriman
pipmerriman@btinternet.com
http://www.lancaster-
beekeepers.org.uk

Manchester & District
Joy Jackson
joy.jackson@mdbka.com
http://www.mdbka.com

Sedbergh & District BKA (Cum-
bria & Yorkshire Dales)
Jane Callus-Whitton
janeecwhitton@yahoo.co.uk
http://sedberghanddistrict-
beekeepers.webs.com

**Institute of Northern
Ireland Beekeepers**
Caroline Thomson
secretary@inibeekeepers.
com
http://www.inibeekeepers.
com

Berkshire Beekeepers
Martin Moore
martinmoore@uwclub.net
http://www.berkshirebee-
keepers.btik.com

Buckinghamshire
Dr B Cullen
beulah.cullen@virgin.net
http://www.buckscountybee-
keepers.co.uk

Dover & District
Mrs M Harrowell
the.harrowells@btinternet.
com http://www.ddbka.com

Hampshire
Zoe Semmens
secretary@hampshirebee-
keepers.org.uk
http://www.hampshirebee-
keepers.org.uk

Kent
Mrs J D Spon-Smith
jennifer@spon-smith.com
http://www.kentbee.com

Medway
Sheila Stunell
sheila.stunell1@btinternet.
com
http://www.mbka-news.com

Newbury & District
Virginia Arnott
v.arnott@ntlworld.com
http://www.newbury
beekeepers.org.uk

Oxfordshire
Michael Williams
secretary@obka.org
http://oxfordshire
beekeepers.com

Surrey
Sandra Rickwood
secretary@surrey
beekeepers.org.uk
http://www.surrey
beekeepers.org.uk

Sussex
Liz Twyford
secretary@sussexbee.org.uk
http://www.sussexbee.org.uk

Sussex West
Graham Elliott
grahammt@tiscali.co.uk
http://www.bbka.org.uk/
local/westsussex/

Thanet
Mrs Rowena Pearce
pearceinsummerfield@tiscali.
co.uk
http://thanetbeekeepers.
org.uk

Vale & Downland
Mrs Jane Greenhalgh
jane.greenhalgh@lineone.net
http://www.valeanddown-
landbees.org.uk

Avon
Rosemary Taylor
rosemary.taylor@tiscali.co.uk
http://www.avonbeekeepers.
co.uk

Bournemouth & Dorset South
Peter Darley
peter.darley@talktalk.net
http://www.bads-bka.org

Cornwall
Mrs J Cooper
julia.i.cooper@btinternet.com
http://www.cbka.co.uk

Cornwall West
Kate Bowyer
secretary@westcornwallbka.
org.uk
http://www.westcornwallbka.
org.uk

Devon
Colin Sherwood
c.j.sherwood@btinternet.com
http://www.devonbeekeep-
ers.org.uk

Dorset
Liz Rescorla winkyozz@
tiscali.co.uk http://www.
bbka.org.uk/local/dorset/

Gloucestershire
Ruby Savage
ruby.savage17@gmail.com
http://www.gbka-cg.org.uk

Somerset
Dr. Richard Bache
secretary@somersetbee-
keepers.org.uk
http://www.somersetbee-
keepers.org.uk

Wiltshire
Claire Barker
clairebarker@outlook.com
http://www.wiltshirebee-
keepers.org.uk

Jersey
Judy Collins
judybees@collinsje.net
http://www.bbka.org.uk/local/
jersey/

Herefordshire
Wendy Cummins
enquiries@herefordshire
beekeepers.org.uk
http://herefordshire
beekeepers.org.uk

Ludlow & District
Alan Jenyon
alan.jenyon@gmail.com
http://www.ludlowbee
keepers.org.uk

Shropshire
Chris Currier
chris.currier@btopenworld.
com
http://www.shropshirebees.
co.uk

Shropshire North
Nigel Hine
nahine@btinternet.com
http://www.nsbka.co.uk

Staffordshire North
David Teasdale
secretary@northstaffsbees.
org.uk
http://www.northstaffsbees.
org.uk

Staffordshire South
Mrs Lynne Lacey
lynne.lacey123@btinternet.
com
http://southstaffsbeekeepers.
co.uk

BBKA

Stratford Upon Avon
Mike Osborne
beekeepers@stratford-upon-avon.freeserve.co.uk
http://www.stratfordbee-keepers.org.uk

Warwickshire
Gill Rose
warwickshirebkasecretary@aol.com
http://www.warwickshirebee-keepers.org.uk

Worcestershire
Chris Broad
chrisbroad1964@btinternet.com
http://www.wbka.net

Wye Valley
Susan Quigley
quigley.susan@hotmail.co.uk http://www.bbka.org.uk/local/wyevalley/

Yorkshire
Roger Chappel
secretary@ybka.org.uk
http://ybka.org.uk

ASSOCIATION EXAMINATION SECRETARIES

AVON, Position Vacant
Please contact
Hon. General Secretary
Julie Young
01179 372 156
BERKSHIRE,
Mrs Rosemary Bayliss
Norbury, Coppid Beech Hill,
Binfield, Berkshire.
RG42 4BS
01344 421747
BOURNEMOUTH, Mrs. M. Davies
57 Leybourne Avenue
Ensbury Park
Bournemouth
Dorset BH10 6ES
01202 526077

BUCKS, John Chudley
Orchard Lea, Oxford Street
Lea Common
Great Missenden HP16 9JT
01494 837544
jlchudley@tiscali.co.uk
CHESHIRE
Graham Royle NDB,
7, Symondley Road,
Sutton,
Macclesfield. SK11 0HT
01260 252 042
CORNWALL
Mrs. Susan Malcolm
Fig Tree, 333 New Road
Saltash, Cornwall
PL12 6HL
01752 845496

DEVON, Roger Lacey
Gatchell House
Toadpit Lane, Ottery St Mary
Devon EX11 1TR
01404 811733
devonbees@pobox.com
DORSET, K.G.Bishop
72 Alexandra Road
Bridport DT6 5AL
01308 425479
DURHAM, G. Eames
23 Lancashire Drive
Belmont, Durham,
DH1 2DE
01913 845220
george.eames@durham.ac.uk

ESSEX, Pat Allan
8, Frank's Cottages
St. Mary's Lane
Upminster, RM14 3NU
pat.allen@btconnect.com
GLOUCESTERSHIRE
Bernard Danvers
120a Ruspidge Road
Cinderford,
Gloucestershire
GL143AG
01594 825063
GWENT, Mrs J Bromley
Ty Hir, Monmouth Road
Raglan, Usk. NP15 2ET
01291 690331
bromleyjan@hotmail.com
HAMPSHIRE, Mrs Peggy Mason
37 Springford Crescent
Lordswood,
SO16 5LF
023 8077 7705
H'GATE & RIPON, Peter Ross
The Wheelhouse, The Green,
Old Scriven, Knaresborough
HG5 9EA, 01423 866565,
pjeross@btinternet.com
HEREFORDSHIRE, Len J. Dixon
The Square, Titley,
Kington
Herefordshire HR5 3RG
01544 230884
beeline2ljd@yahoo.co.uk
HERTFORDSHIRE,
R. E. A. Dartington
15 Benslow Lane
Hitchin SG4 9RE
01462 450707
gray.dartington@dial.pipex.
com

ISLE OF WIGHT, Mrs M. Case
Limerstone Farm,
Limerstone, Newport,
Isle of Wight, PO30 4AB
01983 740223
gcase90337@aol.com
KENT, P. F. W. Hutton
22 Good Station Road
Tunbridge Wells,
TN1 2DB
01892 530688
LANCASHIRE & NW
Edward Hill
3 Sandy Lane, Aughton
Ormskirk
L39 6SL
01695 423137
LEICESTERSHIRE & RUTLAND
Brian Cramp
2 Woodland Drive, Groby
Leicester
LE6 0BQ
01162 876879
LINCOLNSHIRE, R. J. B. Hickling
Linden Lea, Sandbraes Lane,
Caistor, LN7 6SB
01472 851473
MIDDLESEX
Mrs Jo V Telfer
Midwood House
Elm Park Road, Pinner
Middlesex HA5 3LH
020 8868 3494
e-mail, jvtelfer@hotmail.
com
NOTTINGHAMSHIRE
Dr Glyn D Flowerdew
Knight Cross Cottage
Newstead Abbey Park
Ravenshead
Nottinghamshire NG15 8GE
01623 792812

OXFORDSHIRE, Terry. Thomas
4 Kirk Close
Oxford, OX2 8JN
01865 558679
PETERBOROUGH, P. G .Newton
65 Queen Street, Yaxley
Peterborough PE7 3JE 01733
243349
SHROPSHIRE NORTH
Paul Curtis
1 Hammer Close
Overton-on-Dee, Wrexham
Clwyd LL13 0LD0
01691 624296
SOMERSET, Mrs Angela Bache
Greenway House
Badgers Cross
Somerton TA11 7JB
 Tel 01458 273149
STAFFS.Nth Dr. Nick C Mawby
Glenwood, Wood Lane
Longsdon,
Stoke on Trent ST9 9QB
01538 387506
info@northstaffsbees.org.uk
STAFFS. SOUTH
Tony Burton
96 Weeping Cross, Stafford,
Staffordshire. ST9 9QB
01538 399322
SUFFOLK, Mr Ian McQueen
643 Foxhall Road, Ipswich,
Suffolk, IP3 8NE
01473 420187
SURREY, Mrs. A. Gill
143 Smallfield Road
Horley, RH6 9LR
01293 784161

BBKA

SUSSEX, Nigel Champion
45 Ridgeway,
Hurst Green
Etchingham
East Sussex TN19 7PJ
01580 860379
SUSSEX WEST
Mrs A. S. Gibson-Poole
Mont Dore, West Hill
High Salvington
Worthing, BN13 3BZ
01903 260914
TWICKENHAM, Chris Deaves
12 Chatsworth Crescent
Hounslow,
Middlesex
TW3 2PB
0208 5682869
e-mail,
c-deavs@compuserve.com

WARWICKSHIRE, P.D. Lishman
Aston Farm House
Newtown Lane
Shustoke, ColeshillB46 2SD
01676 540411
WILTSHIRE, John Troke
The Lythe
Hop Gardens
Whiteparish, Salisbury,
Wiltshire SP5 2SS
01373 822892

WORCESTERSHIRE, D.P. Friel
17 Tennal Rd, Harborne
Birmingham, B32 2JD
0121 427 1211
YORKSHIRE, Brian Latham
Tel: 0113 264 3436
Mob: 07765842766
brian.latham@ntlworld.com

Where Associations have no Examinations Secretary the Association Secretary deals with examinations. To help future candidates it is suggested that Associations without an Examination Secretary appoint one. Associations are responsible for arranging a suitable room for the written examinations and recommending an invigilator.

If you live in an area without a nominated Exam Secretary, you should contact Mrs Val Frances, 39 Beevor Lane, Gawber, Barnsley, S75 2RP   Tel 01226 286341. e-mail, valfrances@blueyonder.co.uk

HOLDERS OF THE BBKA SENIOR JUDGES CERTIFICATE

ASHLEY, Mr. T. E.
Meadow Cottage
Elton Lane, Winterley
Sandbach
Cheshire CW11 4TN
BADGER, M.J , MBE
14 Thorn Lane
Leeds, LS8 1NN
BLACKBURN, Mrs. H.M
15 Highdown Hill Road
Emmer Green
Reading RG4 8QR
BROWN, Mrs. V
BUCKLE, M.J
The Little House
Newton Blossomville
Bedford MK43 8AS
01234 881262
martin@newtonbee.fsnet.
co.uk
CAPENER, Rev. H.F.
1 Baldric Road
Folkestone CT20 2NR
COLLINS, G.M. , NDB
72 Tatenhill Gardens
Doncaster DN4 6TL
COOPER, Miss R.M
10 Gaskells End
Tokers Green
Reading RG4 9EW
DAVIES, Mrs. M
57 Leybourne Avenue
Ensbury Park
Bournemouth BH10 6HE

DIAPER, B
B Diaper
57 Marfield Close
Walmley
West Midlands
0121 313 3112
or 07711 456932
DICKSON, Ms. F
Didlington Manor
Didlington, Thetford
Norfolk IP26 5AT
DUFFIN, J.M
Upper Hurst
Salisbury Road, Blashford
Ringwood
Hampshire BH24 3PB
01425 474552
FIELDING, L.G
Linley, Station Road
Lichfield WS13 6HZ
MacGIOLLA CODA, M.C.
Glengarra Wood, Burncourt
Cahir, Co. Tipperary
Republic of Ireland
McCORMICK, E.
14 Akers Lane, Eccleston St.
Helens, Lancs WA10 4QL
MOXON, G
9 Savery Street
Southcoates Lane
Hull HU9 3BG

ORTON J
Occupation Road, Sibson
Nuneaton CV13 6LD
SALTER T.A , MBE
44 Edward Road, Clevedon
North Somerset BS21 7DT
SYMES, C.J
189 Marlow Bottom Road
Marlow SL7 3PL
TAYLOR, A.J
The Old Pyke Cottage
Hethelpit Cross, Staunton
GL19 3QJ
WILLIAMS, M
Tincurry, Cahir,
Co Tipperary, Eire
YOUNG, M
Mileaway, Carnreagh Hills-
borough,
Northern Island BT26 6LJ

BIBBA

BEE IMPROVEMENT & BEE BREEDERS' ASSOCIATION

www.bibba.com

SECRETARY
Roger Cullum-Kenyon
The Green
Galmpton
Kingsbridge
South Devon
TQ7 3EY
01548 560021
cullumkenyon@
btopenworld.com

MEMBERSHIP SECRETARY
Iain Harley
93 Dunsberry
Bretton
Peterborough
Cambridgeshire
PE3 8LB
01733 700740
iain.harley42@ntlworld.com

SALES SECRETARY
John Hendrie
26 Coldharbour Lane
Hildenborough
Tonbridge
Kent
TN11 9JT
sales@bibba.com

BIBBA is an organisation devoted to encouraging beekeepers to breed native or near native bees. The bees more suited to our environmental circumstances than other sub species.

BIBBA's aims are publicised through their magazine 'Bee Improvement', books, workshops, lectures and conferences.

BIBBA co-operates worldwide with Beekeeping and breeding groups interested in conserving and improving their own native bees.

Breeding techniques advocated include:

• Assessment of colonies by observation, recording certain criteria on standard record cards.

• Determination and purity of sub species by measurement of morphometric characters and mitrochondial DNA.

• Use of mini nucs for the mating of queens economically

BIBBA Publications include:

• The Honeybees of the British Isles by Beowulf Cooper

• Breeding Techniques and Selection for Breeding of the Honeybee by Prof. F. Ruttner

• The Dark European Honey Bee by Prof. F. Ruttner, Rev. Eric Milner and John Dews

• Breeding Better Bees using Simple Modern Methods by John E. Dews and Rev.Eric Milner

• Better Beginnings for Beekeepers by Adrian Waring - second edition.

BIBBA encourages the formation of Bee Breeding Groups, and the sharing of knowledge between groups by the provision of genetic material.

Look out for Bee Improvement days and Queen Rearing events in the bee press and on www.bibba.com.

THE C.B. DENNIS BRITISH BEEKEEPERS' RESEARCH TRUST

REGISTERED CHARITY NO. 328685

Aims

This independent Charitable Trust awards grants for research at Universities and institutions on the basis of scientific merit and supports young bee scientists by providing funding for studentships and training. Since its inception the Trust has funded work on a wide range of topics related to both honey bees and other bees. Recent projects have focused on gaining a better understanding of the relationship between Varroa destructor and deformed wing virus (DWV) of honey bees and the development of strategies to moderate their damaging effects. Currently the Trust is supporting a number of young scientists who are investigating land use and honey bee bacterial associations, bee pollination in an agricultural landscape and beekeeping and conservation.

Awards

The Trust is administered by a group of seven Trustees all of whom are, or have been career scientists. They therefore have first-hand knowledge of both writing and evaluating research proposals and several have extensive practical experience of working with bees in a professional or hobbyist capacity. This expertise ensures that work funded by the Trust is properly evaluated and provides the greatest possible advantage for bees. Meetings are held twice a year in April and October to evaluate submitted research applications.

Donations

The Trust is pleased to acknowledge the loyal support it already receives from several local beekeeping associations and many individuals. All donations, however small, will be added to the invested capital and bee research in Britain will benefit from the income in perpetuity. Full details of the activities of the Trust, outputs of the research funded and grant application forms can be obtained from www.cbdennistrust.org.uk

CABK

✉ ☎

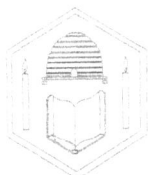

THE CENTRAL ASSOCIATION OF BEEKEEPERS

www.cabk.org.uk

SECRETARY, Pat Allen
8 Frank's Cottages
St Mary's Lane
Upminster, RM14 3NU
pat.allen@btconnect.com

PRESIDENT, Prof. R.S. Pickard
pickard.r@btopenworld.com

TREASURER, Jim Vivian-Griffiths,
Catkins, 17 The Kymin,
Monmouth, Monmouthshire
NP25 3SF
jimvg45@gmail.com

PROGRAMME SECRETARY
Pam Hunter
Burnthouse
Burnthouse Lane
Cowfold, Horsham
RH13 8DH
pamhunter@burnthouse.org.uk

EDITOR, Pat Allen
8, Frank's Cottages
St. Mary's Lane
Upminster, RM14 3NU

SALES AND DISTRIBUTION,
Bill Fisher
The Old Farmhouse,
Farm Road, Chorleywood,
Hertfordshire
WD3 5QB
theoldfish@hotmail.com
07973 626464

The Central Association of Beekeepers in its present form dates from the time of the reorganisation of the British Beekeepers' Association in 1945. The BBKA was originally made up of private members only. However as County Associations were formed they applied for affiliation and were later permitted to send delegates to meetings of the Central Association, as the private members were then known. This arrangement became unsatisfactory as the voting power of the Central Association greatly outnumbered that of the County Associations and so in 1945 a new Constitution was drawn up whereby the Council comprised Delegates from the Counties and Specialist Member Associations. The private members then formed themselves into a Specialist Member Association with the designation 'The Central Association of the British Beekeepers' Association'; this was later shortened to its present style.

The Association was able to devote itself to its own particular aims, to promote interest in current thought and findings about beekeeping and aspects of entomology related to honey-bees and other social insects. Lectures given by scientists and other specialists are arranged, printed and circulated to members, as has been done since 1879.

A Spring Meeting with three lectures plus Annual General Meeting is held in London, and an Autumn Weekend Conference in the Midlands. In addition a lecture is given at the Social Evening held during the National Honey Show. Subscriptions are £15 per annum for an individual, £18 for dual membership, £20 for corporate membership.

THE EVA CRANE TRUST

www.evacranetrust.org

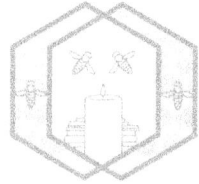

The Trust was formed by Dr Eva Crane herself. It was enhanced by the residue of her estate bequeathed to the Trust on her death in 2007.

Trust Chairman
Richard Jones

Dr Crane's research was meticulous and the recording of information - so that original material could be traced and used by succeeding generations - was a vital part of her work. In her lifetime she had over 300 papers and articles published, and she contributed many learned tomes to the shelves of bee lovers worldwide.

The aim of the Trust is to continue Dr Crane's work in the way she would have liked it to evolve. This includes advancing the understanding of bees and beekeeping by the collection, collation and dissemination of science and research worldwide, as well as recording and propagating a further understanding of beekeeping practices through historical and contemporary discoveries.

The Trust, as well as being Dr Crane's way of ensuring her work continues, is a memorial whereby it may be possible to help fund others who can build on the foundations of sound academic research laid down in her many publications. Grants may be made to individuals and organizations that might otherwise find funding difficult in this specialized field. Applications will be considered from anywhere in the world but must be made in writing in the English language, preferably using the form on the website.

The website, which will be developed and expanded in the coming months, can be found at:

http://www.EvaCraneTrust.org

Similar information can be obtained by writing to:

The Eva Crane Trust, c/o Withy King Solicitors,

5-6 Northumberland Buildings, Bath, BA1 2JE, UK

CONBA

CONBA-UK & Ireland
COUNCIL OF NATIONAL BEEKEEPING ASSOCIATIONS IN THE UNITED KINGDOM and IRELAND

Incorporating the beekeeping organisations of : England, Channel Islands Isle of Man, Wales, Scotland, Ulster, Ireland and The Bee Farmers Association

SECRETARY Terry Gibson
17, Ffolkes Drive, Gaywood, King's Lynn, Norfolk. PE30 3BX
01553 674051
t.gibson1@virgin.net

CHAIRMAN, Mervyn Eddie (UBKA)
3b Old Road
Upper Ballinderry
Lisburn, Co. Antrim. BT28 2NJ
0289 265 2580
eddie_mervyn@yahoo.co.uk

VICE CHAIR Michael Gleeson (FIBKA)
Ballinakill
Enfield
Co.Meath, Ireland.
+353 (0)4695 41433
mgglee@eircom.net

HON.TREASURER Martin Tovey (BBKA)
11, Coach Road
Warton, Carnforth
Lancashire. LA5 9PP
01524 730451
martintovey@hotmail.co.uk

CONBA was established in 1978 to promote the aims and objectives of the national beekeeping associations of England, Scotland, Ulster, Wales and Ireland, and the Bee Farmers Association. Its purpose is to represent the interests of beekeepers' with local, national and international authorities. A representative delegate from each of the member country associations occupies the chair for a period of two years, on a rotational basis.

The council meets twice per year, normally at the Spring Convention and at the National Honey Show in London, with the remaining meeting by rotation in the member association's country. Council business consists of any matters of common interest to all its members. CONBA provides representation of its membership at the European Union (EU) through two specific committees, COPA and COGECA (COPA – Comite des Organisations Professionelles Agricoles de la CEE); (COGECA Comite de la Cooperation Agricole de la CEE); and the Honey Working Party (HWP).

The Honey Working Party meetings are held at Brussels. This committee liases with the European Commission in relation to apicultural matters concerning the member states of the European Union (EU). These matters are subsequently presented to the European Parliament for its consideration, implementation or revision or rejection. The subsequent approval of such matters results in establishing legislation, government support and possible EC funding relating to the practice of apicultural production in the UK through its membership of the EU.

COUNCILLORS REPRESENTING THE MEMBER ASSOCIATIONS
BBKA Tovey, Martin; Routh, Julian
BFA Bancalari, David; Ginman, Margaret
FIBKA Gleeson, Michael; McCabe, Philip
SBA McAnespie, Phil; Wright, Bron SBA
UBKA Eddie, Mervyn; Wright, David
WBKA Sweet, Dinah; Williams, Harold

DEVON APICULTURAL RESEARCH GROUP

CHAIRMAN, Richard Ball
Stoneyford Farmhouse
Colaton Raleigh
Sidmouth, Devon EX10 0HZ
01 395 567 356

HON SECRETARY, Vacant
Contact Chairman

PUBLICATIONS OFFICER, David Loo
25 Woodlands
Newton-St-Cyres, Exeter
Devon EX5 5BP
0139 285 1472

TREASURER, Bob Ogden
Pennymoor Cottage
Pennymoor, Tiverton
Deven EX16 8LJ
01363 866687

All titles cost £2.50 per copy (post free) from the Publications Officer (tel. 01392 851472). Discounts are available for BBKA affiliated Associations **Please contact the Publications Officer for details**

D A R G is an independent group of experienced enthusiastic beekeepers whose primary aim is to collect and analyse data on matters of topical interest which may assist their apicultural education and promote the advancement of beekeeping. At their regular meetings, DARG members discuss various topics in open forum, during which they exchange ideas and information from their personal beekeeping knowledge and experience. They also undertake suitable research projects which further the Group's aims.

TOPICS CURRENTLY BEING UNDERTAKEN

- Use of management (mechanical) methods including shook colonies for varroa control.
- Brood cell size in natural comb.
- A survey of useful bee plants, shrubs and trees in the South West.
- Drone movement between colonies.

In conjunction with Devon BKA

- Survey of Nosema in the County of Devon.
- Survey of drone laying queens in the County of Devon.

PUBLICATIONS AVAILABLE

- **The Beeway Code.** A common sense guide for beginners to help avoid problems with neighbours and produce a safe and peaceful apiary.
- **Seasonal Management**. A useful aid to planning your work effectively
- **Queen Rearing.** Providing detailed help in rearing new queens in order to promote vigorous colonies.
- **The selection of Apiary sites** full of tips for choosing the right sites for your bees.

142

THE FEDERATION OF IRISH BEEKEEPERS' ASSOCIATIONS

http://www.irishbeekeeping.ie

Secretary: Mr Stuart Hayes,
54 Glenvara Park, Knocklyon, Dublin 16
Tel No 085-1602613, email: fibka.secretary@gmail.com

ANNUAL SUMMER COURSE

The 2015 Beekeeping Summer Course will take place at the Franciscan College, Gormanston, Co Meath from Sunday 25th July to Friday 31st of July. The Guest Speaker will be Professor Ingemar Fries at the Department of Ecology, Swedish University of Agricultural Sciences, Uppsala, Sweden.

For further information and to secure your place, contact the Summer Course Convenor
Mr Michael G Gleeson, Ballinakill, Enfield, Co Meath. Tel No 046-9541433/087-6879584, email mgglee@eircom.net or visit http://www.irishbeekeeping.ie/gormanston/gormprog2014.html

PUBLICATIONS:

Having Healthy Honeybees - Published by F.I.B.K.A. Editor John McMullan, Ph.D.
The aim of this book is to help beekeepers establish healthy honeybee colonies, assess their condition and take appropriate action. Diseases are dealt with in a concise format to improve readability and are referenced to the latest peer-reviewed research. The book emphasises the importance of proper set-up, involving an integrated approach to health management – in effect a preventative system that comes at little extra cost to the beekeeper
Cost €15 + P & P of €2 each
Bulk buying available to Associations In packs of 10 or 20 books, available at €12 each + P & P of €10 for packs of 10 or 20.
The recommended price is €15 per copy.
It is highly recommended for those doing the various FIBKA Examinations.
Available from Mr Michael G Gleeson, Ballinakill, Enfield, Co Meath.

OFFICERS:
President: Mr Eamon Magee,
222 Lower Kilmacud Road, Goatstown, Dublin 14.
Tel No 01-2987611
E-mail eamonmagee222@gmail.com

Vice-President:
Mr Gerry Ryan,
Deerpark, Dundrum, Co Tipperary
Tel No 062-71274/087-1300751, E-mail ryansfancy@gmail.com

P R O: Mr Philip McCabe,
"Sherdara",
Beaulieu Cross, Drogheda, Co Louth.
Tel No (041-9836159),
E-mail philipmccabe@eircom.net

Life Vice-Presidents:
Mr P O'Reilly,
11 Our Lady's Place, Naas, Co Kildare
Tel No (045-897568),
E-mail jackieor@indigo.ie

FIBKA

✉ ☎

Mr MI Woulfe,
Railway House,
Midleton, Co Cork
Tel No (021-4631011),
E-mail glenanorehoney@
eircom.net

Mrs Frances Kane,
Firmount, Clane, Co Kildare.
Tel No (087-2450640) or
(045-893150)

Editor: Ms Mary Montaut,
4 Mount Pleasant Villas,
Bray, Co Wicklow.
Tel No 01-2860497. E-mail
mmontaut@iol.ie

Manager: Mr Dermot O'Flaherty,
Rosbeg, Westport,
Co Mayo
Tel No 098-26585/
087-2464045
E-mail:
fibka.manager@anu.ie

Treasurer: Mr Pat Finnegan,
Mullaghmore Road,
Cliffoney, Co Sligo
Tel No 071-9166597/
087-9272692,
email: finnegan@iol.ie

**Education Officer:
Mr John Cunningham,**
Ballygarron, Kilmeaden,
Co Waterford
Tel No 051-399897/086-
8389108,
email: john3cunningham@
hotmail.com

**Summer Course Convenor:
Mr Michael G Gleeson,**
Ballinakill, Enfield,
Co Meath.
Tel No 046-9541433/
087-6879584,
email mgglee@eircom.ne

Tel No (046-9541433) & (087-6879584), E-mail mgglee@ eircom.net

Bees, Hives and Honey - Published by F.I.B.K.A. – Edited by Eddie O'Sullivan
This book has been compiled from writings by some of Ireland's most prominent Beekeepers of the present day. It is an instruction book on beekeeping published as a Millennium project and should prove a modern treatise on the craft of beekeeping and its associated products. There are over 200 pages, also many photographs and illustrations. Price €12.70 (Paperback) or €19 (Hardback) Available from Eddie O'Sullivan, Phone: 021-4542614, Email: eosbee@indigo.ie

The Irish Bee Guide – by Reverend J.D. Digges. First published in 1904, it was proclaimed as an excellent book on beekeeping. It also won a place as a notable production in the literary context. It eventually ran to sixteen editions and sold seventy-six thousand copies overall. The name was changed in the second issue to The Practical Bee Guide.
Now, one hundred years later, a decision has been taken to honour this great work. What better way to do it than to re-issue the book as it was in 1904 when it first entered the literary world. The re-print is an exact replica of the original first edition. The price per copy is Hardback€30 and Softback €20
Available from Eddie O'Sullivan, Phone: 021-4542614, Email: eosbee@indigo.ie

An Beachaire – The Irish Beekeeper the monthly organ of FIBKA, subscription €25.00 (Irish Republic), £25 Stg (Northern Ireland/Great Britain) post free from The Manager
Mr Dermot O'Flaherty, Rosbeg, Westport, Co Mayo Tel No 098-26585/ 087-2464045
E-mail:fibka.manager@anu.ie
Readership of the Journal in Northern Ireland carries third party insurance public liability cover up to €6.500, 000 on any one claim and product liability cover up to €6.500, 000 on any one claim, on payment of £5.00 Stg extra.

FIBKA

LIBRARY

The library is owned and controlled by FIBKA. It contains very many valuable books ancient and modern, available to members for return postage only. The Librarian is Jim Ryan, Innisfail, Kickham Street, Thurles, Co Tipperary. Email jimbee1@eircom.net

EDUCATION

The Federation of Irish Beekeepers' Associations (FIBKA) examination system is run by the Education Officer under the direction of the Examination Board; the Board which is made up of members from the FIBKA and the Ulster Beekeepers' Association (UBKA) is appointed by the Executive Council of the FIBKA.

There are seven levels of examination: Preliminary, Intermediate, Senior, Lecturer and Honey Judge Examinations are held during the Summer Course at Gormanston and Preliminary and Intermediate examinations are also held at Provincial Centres.

The Lecturer's examination takes place in the presence of three Examiners, one of whom is the invited Senior Gormanston Summer Course lecturer and also acts as the Extern Examiner.

The Intermediate Proficiency Apiary Practical Examination, the Practical Beemasters Examination and the Apiary Practical component of the Senior Examination are arranged by the Education Officer and take place in the candidate's own apiary during the beekeeping season and are conducted by two Examiners.

The seven levels of examinations for proficiency certificates and their eligibility requirements are as follows:

Preliminary:

For beginners - no prerequisites.
Intermediate:
The Preliminary Certificate of the FIBKA or the BBKA Basic Certificate must be held for at least one year.

Senior:

Intermediate Certificate and at least five years beekeeping experience.
Intermediate Proficiency Apiary Practical
The Intermediate Proficiency Apiary Practical Examination is intended to be part of a stream that will lead to the Practical Beemasters Certificate. The examination is designed to be less "academic" and there are no written examination papers; (it is not part of the Intermediate Certificate Examination).

The examination will take place in the candidate's own apiary and the Examiners will be two Federation Lecturers appointed by the Executive Council. The pass mark is 70%. 20% of the marks scored may be carried forward to the Practical Beemasters Examination

The prerequisites for Intermediate Proficiency Apiary Practical Examination are: the Preliminary Certificate and at least three years' beekeeping experience satisfactory to the Education Board.

The present prerequisites for the Practical Beemasters Certificate are the Preliminary Certificate and at least five years' beekeeping experience satisfactory to the Examination Board - in the future, an additional prerequisite will be the Intermediate Proficiency Apiary Practical Examination.

Practical Beemaster:
Preliminary Certificate and at least five years' beekeeping experience satisfactory to the Examination Board.

Honey Judge:
Intermediate and Practical Beemaster Certificates, successful showing, having obtained a minimum of 200 points at major shows and a record of stewarding under at least four FIBKA Honey Judges.
Lecturer:
Senior Certificate.

Provincial Examinations
Preliminary and Intermediate examinations will be held at provincial centers on the Saturday closest to 6th April (Intermediate) and May 24th (Preliminary). Please note that the minimum number of candidates for a Centre is five for Intermediate and ten for Preliminary. Neighbouring associations may pool their candidates to reach those numbers.
A candidate may sit one Intermediate paper at the Provincial Examination and the other paper at the Summer Course.
The fees for all examinations are valid for the year of application only and are listed on the application forms which may be downloaded from the website. In extreme cases, such as illness (a doctor's certificate must be provided); the examination fee may be held over for one year. There are separate entry forms for the Provincial and Gormanston Summer School Examinations
Fees for Repeat Examinations are the same as for the original examination. Applications to sit the Examinations should be sent to the Education Officer, before the closing dates given above for the Provincial Examinations (applications are however acceptable up to one week after the closing date on payment of a late entry fee which is equal to double the original fee) and before May 1st for the Summer Course Examinations Applications for the Preliminary Examination are also accepted at the Summer Course.

NATIONAL HONEY SHOW

This is held at Gormanston College in conjunction with the annual Beekeeping Course. The Schedule contains 41 Open Classes and 3 Confined classes with €1,000 in prizes. Over 30 Challenge Cups and Trophies are presented for the competition.

Honey Show Secretary: Mr Graham Hall, "Weston", 38 Elton Park, Sandycove, Co Dublin. Tel No (01-2803053) & (087-2406198), E-mail GrahamHall@iolfree.ie

INSURANCE

The limit of indemnity of public liability policy is €6.500, 000 arising from one accident or series of accidents. There is also product liability of €6.500, 000 arising from any one claim. The policy extends to all registered affiliated members whose subscriptions are fully paid up on the 31st December of any one year and whose names are entered in the FIBKA register held by the Treasurer.

FIBKA

✉ ☎

ASSOCIATION SECRETARIES

ARMAGH & MONAGH
Mrs. Joanna McGlaughlin
26 Leck Road,
Stewartstown Co Tyrone
BT71 5LS
Tel No 048-87738702/077-
68107984.
secretary@ambka.org

Ashford
Ms Michele O'Connor,
087 2505205
info@wicklowbees.com

Ballyhaunis
Mr Gerry O'Neill,
Drimineen South,
Knock Road, Claremorris,
Co Mayo.
Tel No 087 2553533
ballyhaunisbeekeepers@
gmail.com

Banner
Mr Frank Considine,
Clohanmore Cree,
Kilrush, Co Clare,
Tel No 087-6740462,
bannerbees@gmail.com ,

Beaufort
Mr Padruig O'Sullivan,
Beaufort Bar & Restaurant,
Beaufort, Co Kerry.
Tel No 087-258993006,
beaurest@eircom.net

Carbery
Mr Sean O'Donovan,
Drominidy, Drimoleague,
Co Cork.
Tel No 087-7715001.
seanodonovan10@gmail.
com

Co Cavan
Mr Alan Brady,
Shanakiel House,
Drumnagran, Tullyvin,
Co Cavan
Tel No 086-8127920
alan@alanbrady.ie or Info@
alanbradyelectrical.com

Co Cork
Mr Robert McCutcheon,
Clancoolemore, Bandon,
Co Cork.
Tel No 023-8841714.
bob@cocorkbka.org

Co Donegal
Mr Dan Thompson,
Highfield, Loughnagin,
Letterkenny, Co Donegal
Tel No 074-9125894
dthompson@eircom.net

Co Dublin
Mr Liam McGarry,
24 Quinn's Road,
Shankill, Co. Dublin
Tel No 087 2643492.
mcgarryliam@gmail.com

Co Galway
Dr Anna Jeffrey Gibson,
Ballyclery, Kinvara,
Co Galway
secretary@
galwaybeekeepers.com

Co Kerry
Mr Ruary Rudd,
Westgate, Waterville,
Co Kerry.
Tel No 066-9474251.
rrudd@eircom.net

Co Limerick
Mr Gus McCoy,
Mount Catherine Clonlara
Co. Clare
Tel No 087 1390039 :
gusmccoy1@eircom.net

Co Louth
Mr Tom Shaw,
201 Ard Easmuinn, Dundalk,
Co Louth
Tel No 042-9339619/
086-2361286,
tshaw@iol.ie

Co Longford
Mr Joe McEntegart,
Cleanrath, Aughnacliffe,
Co Longford.
Tel No 087-2481340.
josephmcentegart@yahoo.
com

Co Mayo
Ms Helen Thompson,
Graffy, Killasser,
Swinford, Co. Mayo.
Tel No 087-7584835
info@mayobeekeepers.com
or helen.mmooney@gmail.
com

Co Offaly
Mrs Geraldine Byrne,
4 Sheena, Charleville Rd,
Tullamore, Co Offaly
Tel 086-3464545,
loureiro.byrne@gmail.com

Co Waterford
Ms Colette O'Connell,
4 Davis Street, Dungarvan,
Co Waterford
Tel No 058-41910,
coletteoconnell@ymail.com

Co Wexford
Mr John Cloney,
Ballymotey Beg,
Enniscorthy, Co.Wexford.
Tel No 087 9801015
countywexfordbeekeepers@
gmail.com

Chorca Dhuibhne
Ms Juli Ni Mhaoileoin,
Burnham, Dingle,
Co Kerry
Tel No 086-8337733,
julimaloneconnolly@
gmail.com

Chonamara
Mr Billy Gilmore,
Maam West, Leenane,
Co. Galway
Tel No 091-571183/087-
7942028,
b.gilmore@
connemarabeekeepers.ie

Digges & Dist
Mr Walter Sharpley,
Aughayoula, Ballinamore
Co. Leitrim
Tel No 086 1236207
waltersharpley@gmail.com

Duhallow
Mr Andrew Bourke,
Pallas, Lombardstown,
Mallow, Co Cork
Tel No 087-2783807.
bourke.andy@gmail.com

Dunamaise
Mr Thomas Hussey,
Glenside Portlaoise,
Co. Laois
thomasjhussey@eircom.net

Dunmanway
Elke Hasner,
Kilnarovanagh, Toames,
Macroom, Co.Cork
026 46312/ 087 2525771
elkehasner@gmail.com

East Cork
Mrs Bridie Terry,
Ait na Greine, Coolbay,
Cloyne, Co Cork.
Tel No 021-4652141.
aitnagreine@gmail.com

East Waterford
Mr Michael Hughes,
51 Woodlawn Grove, Cork
Road, Waterford
Tel No 051-373461.
waterfordbees@gmail.com

Fingal
Mr John McMullan,
34 Ard na Mara Crescent,
Malahide, Co Dublin
Tel No 01-8450193.
jmcmullan@eircom.net

Foyle
Mr Martin Coleman,
Greencastle,
Co. Donegal
foylebeekeepers@
gmail.com

Gorey
C/O President,
Gerard M Williams,
Carrigbeg,
Gorey Co. Wexford
Tel No 053-9421823/086-
3634134 e-mail
geraldandvera@eircom.net

Inishowen
Mr Paddy McDonagh,
Milltown, Carndonagh,
Co Donegal.
Tel No 074-9374881.
paddymcdonough@eircom.
net

Iveragh
Mr Shannon
Ware, 4 Ballinskelligs
Holiday Homes,
Ballinskelligs,
County Kerry.
Tel No: 083-3862345
research@gamelab.ca

Killorglin
Mr Declan Evans,
Reeks View Lodge,
Killorglin, Co Kerry.
Tel: 087 175 4078, :
declanjevans@gmail.com

Kilternan
Ms Mary Montaut,
4 Mount Pleasant Villas,
Bray, Co Wicklow.
Tel No 01-2860497.
mmontaut@iol.ie

Mid-Kilkenny
Jer Keohane,
Jenkinstown Park, Kilkenny.
Tel. 056-7767195 / 087-
2523265 /
jkeohane@iece.ie

New Ross
Mr Seamus Kennedy,
Churchtown,
Feathard-on Sea,
New Ross, Wexford
Tel No 051-397259/
086- 3204236.
seamus.kennedy@yahoo.
co.uk

North Cork
Mr. Eamon Nelligan
3 Carriagroghera
Fermoy Co. Cork.
ciaranneligan@gmail.com

North Kildare
Mr Norman Camier,
34 Lansdowne Park,
Templeogue, Dublin 16.
Tel No 01-4932977/
087-2848938,
norman.camier@gmail.com

Nth Tipperary
Mr Jim Ryan,
"Innisfail", Kickham Street,
Thurles, Co Tipperary.
Tel No 0504-22228.
jimbee1@eircom.net

FIBKA

Roundwood
Mr John Coleman,
Hillside Cottage,
Roundhill Haven, Clara
Beg, Roundwood, Co.
Wicklow Tel No 087-795
4385, colemanjkc@gmail.
com

Sliabh Luachra
Mr Billy O'Rourke,
Dooneen, Castleisland,
Co Kerry
Tel No 066-7141870,
siobhancorourke@eircom.
net

Sligo/Leitrim
Mr Peter Carter,
Doon West, Gurteen,
Co.Sligo
slbasecretary@gmail.com

Sneem
Mr Frank Wallace,
Boolananave, Sneem.
County Kerry.
Tel No 086 3522205,
franksneem@hotmail.com

South Donegal
Mr Derek Byrne,
Carrick West, Laghey,
Co Donegal.
Tel No 074-9722340.
dcbyrne@eircom.ie

South Kildare
Mr Liam Nolan,
Newtown, Bagnelstown,
Co Carlow.
Tel No 059-9727281.
liamnolannt@gmail.com

Sth Kilkenny
Mr John Langton,
Coolrainey,
Graiguemanagh,
Co Kilkenny
Tel No 086-1089652,
jjlangton@eircom.net

Sth Tipperary
Mr P J Fegan,
Tickinor, Clonmel,
Co Tipperary.
Tel No 086 1089652,
feganpj@eircom.net

Sth West Cork
Ms Gobnait O'Donovan,
38 McCurtain Hill,
Clonakilty, Co Cork.
Tel No 023-
8833416/083-3069797
gobnaitodonovan@gmail.
com

Sth Wexford
Mr. Dermot O'Grady,
Linden House, Horetown
North, Foulksmills, Co.
Wexford Tel: (051) 565651,
dermaloid@gmail.com

Suck Valley
Ms Anne Towers,
Doonwood,
Mount Bellew, Co Galway.
Tel No 0909-684547/
087-6305714,
annevtravers@gmail.com

The Kingdom
Ms Rebecca Coffey,
75 Ashgrove, Tralee,
Co Kerry
Tel No 066- 7169554,
bexk8@yahoo.co.uk

The Royal Co
Ms Geraldine McCann,
Mooretown, Ratoath,
Co. Meath.
geraldine.toole@ucd.ie

The Tribes
Mr Eoghan O'Riordan,
28 Arbutus Avenue,
Renmore, Galway.
Tel No 091-753470/
087-6184132,
landservices@eircom.net

West Cork
Ms Jacqueline Glisson,
Costa Maningi, Derrymihane
East, Castletownbere,
Co Cork. Tel No 086-
3638249,
jglisson@eircom.net

Westport
Mr Dermot O Flaherty,
Rosbeg, Westport,
Co Mayo
Tel No 098 26585/
087-2464045,
info@mayo-westport.com

INTERNATIONAL BEE RESEARCH ASSOCIATION WEB http://www.ibra.org.uk

CORRESPONDENCE TO:
OPERATIONS DIRECTOR,
Julian Rees
SCIENTIFIC DIRECTOR,
Norman Carreck

Unit 6,
Centre Court,
Main Avenue,
Treforest,
CF37 5YR
029 2037 2409
No fax
www.ibra.org.uk
mail@ibra.org.uk

IBRA - International Bee Research Association promotes the value of bees by providing information on bee science and beekeeping. This charity was founded in 1949 and is supported by members from around the world. IBRA owns one of the largest international collections of bee books and journals, as well as the Eva Crane / IBRA historical collection and a photographic collection. It operates an online bookshop, publishes its own books and information leaflets, as well as scientific journals.

PUBLICATIONS
JOURNAL OF APICULTURAL RESEARCH
A peer reviewed scientific journal that's worldwide and world class. This quarterly publication contains the latest high quality original research from around the world, covering aspects of biology, ecology, natural history and culture of all types of bees.

BEE WORLD
The flagship publication for IBRA members, this quarterly international journal provides a world view on bees and beekeeping. It covers all topics from bee history to the latest findings in bee science.

BEE WORLD PROJECT
The Beeworld project is the name given to IBRA's outreach program, a project that promotes bee education with all ages of school children along with the many individuals, communities, allotment societies and conservationist groups that have an interest in bees.

IBRA

✉ ☎

This initiative encourages teachers, schools and communities to gain a better understanding of bees, their importance to people, medicine, habitat and the planet. The program looks at the problems facing bees within the UK and across the world then searches for solutions through collaborative working.

www.ibrabeeworldproject.com

IBRA BOOKSHOP

The bookshop is accessible via the web site. To support our charitable status IBRA sells a wide range of publications at competitive prices as well as posters, gifts, DVD's and sundries. IBRA is also a publishing house and offers its members a reduction on IBRA products.

MEMBERSHIP

IBRA is proud of its international status and this is reflected by its members who join from all over the world. The membership package now offers more value than ever before: quarterly issues of Bee World, a discount on IBRA publications and online access to a growing back catalogue. For other benefits and the latest information please visit the web site.

Information about all IBRA publications and services can be found via our web site: www.ibra.org.uk

Available and published by IBRA but also available from Northern Bee Books at www.northernbeebooks.co.uk

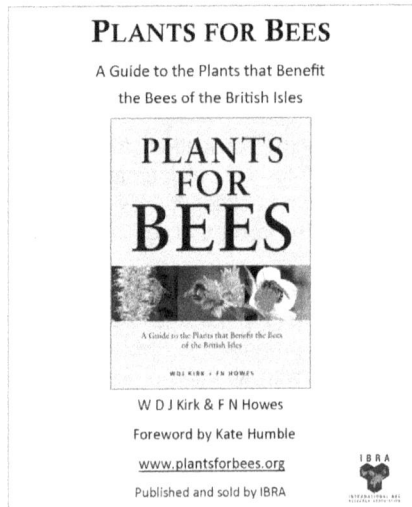

PLANTS FOR BEES

A Guide to the Plants that Benefit the Bees of the British Isles

PLANTS FOR BEES

A Guide to the Plants that Benefit the Bees of the British Isles

W D J Kirk & F N Howes

W D J Kirk & F N Howes

Foreword by Kate Humble

www.plantsforbees.org

Published and sold by IBRA

THE INSTITUTE OF NORTHERN IRELAND BEEKEEPERS (INIB)

www.inibeekeepers.com

Annual Conference and Honey Show. 27th September 2014
Speakers: David Tarpy, Val Francis and Jim Loughery
Lough Neagh Discovery Discovery centre, Oxford Island, Lurgan, BT66 6NJ

Objectives of the Institute

The Institute is established to advance the service of apiculture and to promote and foster the education of the people of Northern Ireland and surrounding environs without distinction of age, gender, disability, sexual orientation, nationality, ethnic identity, political or religious opinion, by associating the statutory authorities, community and voluntary organisations and the inhabitants in a common effort to advance education, and in particular:

- to raise awareness amongst the beneficiaries about bees, bee-keeping and methods of management;
- to foster an atmosphere of mutual support among bee-keepers and to encourage the
- sharing of information and provision of helpful assistance amongst each other.

Affiliation

INIB is affiliated to the British Beekeepers Association.
With 21,100 members the British Beekeepers Association (BBKA) is the leading organisation representing beekeepers within the UK.
As an INIB member, affiliation gives the following benefits.

- BBKA News
- Public Liability Insurance
- Product Liability Insurance
- Bee Disease Insurance available
- Free Information Leaflets to Download
- Members Password Protected Area and Discussion Forum
- Correspondence Courses
- Examination and Assessment Programme
- Telephone Information
- Research Support
- Legal advice
- Representation and lobbying of Government, EU and official bodies.

Events

The Institute holds an annual conference and honey show. The Institute brings to Northern Ireland world renowned expert speakers from USA and Europe to give talks to beekeepers on the latest research and up to date beekeeping methods.

Education

Demonstrations on various topics such as mead making, preparing honey for shows are held during the year.
Courses for honey judges are available.

Honey Bees On Line Studies

INIB has a strong relationship with Professor Jurgen Tautz's of BEEgrouup Biozentrum Universitaet Wuerzburg and his Honey Bee On Line Studies project which continues to develop.

MEMBERSHIP SECRETARY
Lyndon Wortley
Teemore Grange
224 Marlacoo Rd,
Portadown,
BT62 3TD
Membershipsecretary@
inibeekeepers.com

CHAIRMAN
Michael Young MBE
101 Carnreagh,
Hillsborough
BT26 6LJ
02892689724
chairman@
inibeekeepers.com

Holders of the Institute of Northern Ireland Beekeepers Honey Judge Certificate

001.	MICHAEL BADGER MBE	01132 945879	buzz.buzz@ntlworld.com
002.	GAIL ORR	02892 638363	gail.orr@belfasttrust.hscni.net
003.	CECIL MCMULLAN	02892 638675	Madeline.mcmullan@hotmail.co.uK
004.	HUGH MCBRIDE	02825 640872	lorraine.mcbride@care4free.net
005.	LORRAINE MC BRIDE	02825 640872	lorraine.mcbride@care4free.net
006.	BILLY DOUGLAS	02897 562926	
007.	MICHAEL YOUNG MBE	02892 689724	chairman@ inibeekeepers.com
008.	FRANCIS CAPENER	01303 254579	francis@honeyshow.freeserve.co.uk
009.	MARGARET DAVIES	01202 526077	marg@jdavies.freeserve.co.uk
010.	IAN CRAIG	01505 322684	ian'at'iancraig.wanadoo.co.uk
011.	DINAH SWEET	02920 756483	
013.	LESLIE M WEBSTER	01466 771351	leswebster@microgram.co.uk
014.	REDMOND WILLIAMS	003535242617	emwilliams@eircom.net
015.	TERRY ASHLEY	01270 760757	terry.ashley@fera.gsi.gov.uk
016.	IVOR FLATMAN	01924 257089	ivorflatman@supanet.com
017.	ALAN WOODWARD	01302 868169	janet.woodward@virgin.net
018.	DENNIS ATKINSON	01995 602058	dhmatkinson@tesco.net
019	LEO MCGUINNESS	028711 811043	pmcguinness@glendermott.com
020	TOM CANNING		tjcanning@btinternet.com
023	ALAN BROWN	01977 776193	alanhoneybees4u@talktalk.net
024	DAVID SHANNON	01302772837	dave_aca@tiscali.co.uk

USA
021	ROBERT BREWER		rbrewer@arches.uga.edu
022	BOB COLE		
023	ANN HARMAN		

LABRATORY OF APICULTURE & SOCIAL INSECTS (LASI)

UNIVERSITY OF SUSSEX

FURTHER INFORMATION CONTACT
Francis L. W. Ratnieks,
Professor of Apiculture
Laboratory of Apiculture &
Social Insects (LASI)
Department of Biological &
Environmental Science
University of Sussex, Falmer,
Brighton BN1 9QG, UK

01273 872954 (landline),
07766270434 (mob)
F.Ratnieks@Sussex.ac.uk
www.sussex.ac.uk/lasi

Youtube:
LASI Bee Research
& Outreach

LASI was founded in 1995 and is headed by Francis Ratnieks, who is the UK's only Professor of Apiculture. Prof. Ratnieks received his training in honey bee biology at Cornell University and the University of California in the USA. Whilst in the USA he was also a part-time commercial beekeeper with up to 180 hives used for almond pollination and comb honey production.

From 1995 to 2007, LASI was based at the University of Sheffield. In 2008 Prof. Ratnieks moved to the University of Sussex, which provided LASI with excellent facilities for honey bee research. There is an integrated lab space and offices sufficient for 13 researchers with an adjoining apiary, garden, equipment shed and workshop. There are further apiaries on the university campus and in the surrounding countryside.

LASI is the largest university-based laboratory studying honey bees in the UK and is set up both to undertake research and to train the next generation of honey bee scientists. Undergraduate students receive lectures on honey bee biology and can also do research projects on honey bee biology in their final year and assist LASI research via summer bursaries. Graduate students take a PhD that focuses in a particular area of research. Postdoctoral researchers can learn new skills to complement the training they received during their PhD.

LASI research focuses on both basic and applied questions in bee biology and beekeeping. Basic research areas include communication, foraging, colony organization, nestmate recognition and guarding, and conflict resolution. Applied research areas include improved beekeeping techniques, studies of bee foraging and the value of different plants for honey bees and other pollinators, crop pollination, practical studies of honey bee diseases and their management, and practical measures for bee conservation. Collectively, the LASI team has 80 years of research experience with honey bees.

As well as research and teaching, LASI places great emphasis on outreach and communication. Each year LASI runs workshops, gives talks and writes many outreach articles so that research results are also transferred to beekeepers, gardeners, farmers, land owners, the media, the general public, and policy makers.

THE NATIONAL DIPLOMA IN BEEKEEPING

NDB
Beekeeping

The Examinations Board for the National Diploma in Beekeeping was set up in 1954 to meet a need for a beekeeping qualification above the level of the highest certificate awarded by the British, Scottish, Welsh and Ulster Associations.

The Diploma Examination, as designed by the Board, was considered to be an appropriate qualification for a County Beekeeping Lecturer or a specialist appointment requiring a high level of academic and practical ability in beekeeping. It is the highest beekeeping qualification recognised in the British Isles and a high percentage of the past and present holders of the Diploma have given distinguished service to beekeeping education at all levels.

Although the post of County Beekeeping Lecturer has now disappeared, this has merely emphasised the need for some beekeepers to face the challenge of this examination and maintain the high level skills and knowledge needed to keep pace with the increased problems facing all beekeepers at the present time.

The Board consists of representatives from a wide range of organisations and from Government Departments and together form an impressive amalgam of expert knowledge in Beekeeping and Education. Although the National Beekeeping Associations are represented on the Board it is entirely independent of them.

Normally the highest certificate of one of the National Associations is a necessary criterion for eligibility to take the Examination for the Diploma which is held in alternate years. The Written Examination is taken in March, and the Practical, in three sections plus a viva-voce is held in later in the same year.

The Board also organises an annual Advanced Beekeeping Course covering various parts of the syllabus that are difficult to cover by independent study. Lasting

HON. SECRETARY
Mrs Margaret Thomas NDB
Tig na Bruaich,
Taybridge Terrace,
Aberfeldy, Perthshire,
PH15 2BS.

CHAIRMAN,
Ivor Davis NDB
91 Brinsea Road,,
Congresbury
BS49 5JJ
07831 379222

a working week, they cover the main sections of the Syllabus and represent the highest level of training available to British Beekeepers at the present time. The outside lecturers are each acknowledged experts in their particular field. In recent years the Board have been privileged to hold their course at the Fera National Bee Unit at Sand Hutton, York.

In addition the Board organize various short courses at locations in the UK on a number of topics. These are advertised in the bee press and the web site.

For further details regarding the Diploma write, enclosing a stamped A4 SAE to the Secretary, or visit our website: http://www.national-diploma-bees.org.uk/

Those who have gained the National Diploma in Beekeeping

Matthew Allan	Celia Davis	* Geoff Ingold	Pat Rich
*Harry Allen	Ivor Davis	George Jenner	*Fred Richards
*Harrison Ashforth	*Alec S.C. Deans	C. F. Jesson	E. Roberts
*John Ashton	Clive De Bruyn	Simon Jones	*Arthur Rolt
Dianne Askquith-Ellis	A.P. Draycott	A.C. Kessel	*Jeff Rounce
David Aston	M. Feeley	W.E.Large	Graham Royle
*John Atkinson	*Barry Fletcher	G.W. Lumsden	J. Ryding
*Miss E.E. Avey	* David Frimston	*Henry Luxton	J.H. Savage
Dan Basterfield	Oonagh Gabriel	A.S. Mcclymont	*Donald Sims
Ken Basterfield	George Gill	J.I. Macgregor	F.G. Smith
Bridget Beattie	*Reg Gove	Ian Mclean	*George Smith
*Brig. H.T. Bell	*Eric Greenwood	Ian A. Maxwell	J.H.F. Smith
R.W. Brooke	Pam Gregory	Paul Metcalf	Robert Smith
Norman Carreck	Anthony R.W. Griffin	J.Mills	*Ken Stevens
*Rosina Clark	* Robert Hammond	*Bernhard Mobus	*J. Swarbrick
Charles Collins	Ben Harden	G. N'Tonga	Margaret Thomas
Gerry Collins	C.A. Harwood	*Peter Oldrieve	Adrian Waring
*Tom Collins	*Leslie Hender	Gillian Partridge	Brian Welch
*Robert Couston	*Alf Hebden	* E.H. Pee	J. Wilbraham
John Cowan	*Ted Hooper MBE	I.E. Perera	
S. J. Cox	Geoff Hopkinson BEM	E.R. Poole	* - deceased
Jim Crundwell	*G. Howatson	Bill Reynolds	
Beulah Cullen			

THE NATIONAL HONEY SHOW

www.honeyshow.co.uk
24TH – 26TH OCTOBER 2015.

This venue is excellent with Free car parking
Just off the M25 junction 11
Rail from Waterloo to Weybridge or Addlestone

The Show itself is a wonderful competitive exhibition of all the products of the bee-hive, coupled with an excellent series of lectures, workshops and a wide variety of trade and educational stands.

HON TREASURER
C S Mence
27 Acacia Grove
New Malden, Surrey
KT3 3BJ

We recommend that you attend all three days, and suggest that you become a member of the Show – just **£12.00** per annum

For further information, please write to the Hon General Secretary, or Email: showsec@zbee.com or visit our website www.honeyshow.co.uk

- The National Honey Show is the premier honey show within the United Kingdom.

- Although it is named the "National Honey Show", it includes a strong international element.

- As well as the competitive content of the Show, there is also a full programme of lectures and workshops.

- In the Sales Hall, all the major traders and educational organisations are present.

- Further information is readily available on the website www.honeyshow.co.uk or from the Hon General Secretary showsec@zbee.com

NIHBS

✉ ☎

The Native Irish
Honey Bee Society
Apis mellifera mellifera

NATIVE IRISH
HONEY BEE SOCIETY

CHAIRPERSON:
MR. PAT DEASY
chairperson@nihbs.org

SECRETARY:
STEFAN BUZOIANU
stefanbuzoianu@gmail.com

TREASURER:
MR. SEÁN Ó FEANNACHTA
treasurer@nihbs.org

PUBLIC RELATIONS OFFICER:
MS. AOIFE NIC GIOLLA CODA
pro@nihbs.org

WEBMASTER:
JONATHAN GETTY
webmaster@nihbs.org
www.nihbs.org

FACEBOOK PAGE
www.facebook.com/
native-irish-honey-bee-society

What is the Native Irish Honey Bee Society?

NIHBS was established in November 2012 by a group of beekeepers who wish to support the various strains of Native Irish Honey Bee (Apis mellifera mellifera) throughout the country. It is a cross border organisation and is open to all. It consists of members and representatives from all corners of the island of Ireland.

Aims and Objectives -

To promote the conservation, study, improvement and re-introduction of Apis mellifera mellifera (Native Irish Honey Bee), throughout the island of Ireland.

- To establish areas of conservation throughout the island for the conservation of the Native Honey Bee.
- To promote formation of bee improvement groups.
- To provide education on bee improvement and to increase public awareness of the native honey bee.
- To act in an advisory capacity to groups and individuals who wish to promote it.
- To co-operate with other beekeeping organisations with similar aims.
- To seek the help of the scientific community and other stake holders in achieving our aims and objectives.

NIHBS - Plans for the future -

Liaise with groups interested in Native Honey Bees
- Will apply for funding -
- Will help to co-ordinate projects -
- Will raise awareness to beekeepers and the public about Native Honey Bees -
- Talks and lectures

Why Join NIHBS?
- Information on beekeeping events around Ireland – North and South
- queen rearing workshops, talks and lectures.
- Information on how to obtain Native Honey Bees
- Conference discounts
- Discounted entrance fees to events run by NIHBS
- Eligibility to schemes coordinated by NIHBS
- A network of beekeepers interested in our native honeybee

1 years membership costs 20 euro or 20 pounds sterling.
A membership form can be downloaded from website and sent to treasurer or payment can be made on website via paypal.

ROTHAMSTED RESEARCH

www.rothamsted.ac.uk

ROTHAMSTED RESEARCH
Department of AgroEcology,
Rothamsted Research
Harpenden,
Hertfordshire.
AL5 2JQ

STAFF
DR ALISON HAUGHTON
DR JASON LIM
DR SAMANTHA COOK
DR TRISH WELLS
JENNY SWAIN
JONATHAN CARRUTHERS
(PHD STUDENT)
STEVE KENNEDY (BEEKEEPER)

The Rothamsted site provides a unique working environment with specialist modern equipment facilitating research on plant and microbial metabolites, molecular biology and synthetic and analytical chemistry. There is an experimental farm for complex field experiments, and there is a suite of glasshouses, controlled environment facilities, an insectary and a state-of-the-art bioimaging suite housing three new electron microscopes and a confocal laser scanning microscope. Experimental design and analysis are backed up by excellent statistical, computing and library support.

BEE BEHAVIOUR AND POLLINATION ECOLOGY

We are investigating the interaction between bees, crops and the agricultural environment. The spatial and temporal foraging behaviour of honey bees and bumble bees within agricultural areas is being compared. Harmonic radar is being used to track flying bees, and other pollinators such as butterflies, to obtain new information about their flight paths, forage ranges, food preferences and orientation mechanisms.

An integrated model for predicting bumblebee population success and pollination services in agro-ecosystems will be developed by Rothamsted and colleagues at the Environment & Sustainability Institute at the University of Exeter and the University of Sussex, and will provide a powerful tool for shaping recommendations for land managers and policy makers for the sustainable spatial management of pollination within arable and horticultural production systems.

Various qualities of different varieties of crops (oilseed rape and short rotation coppice willows) as important resources for bees are being investigated. The nutritional value of the nectars and pollens, effects on bee fitness and behaviour are key areas of interest.

HONEY BEE PATHOLOGY
Rothamsted's research on the natural history and epidemiology of the infections and parasites of bees has had wide international recognition. However, research on honey bee pathology is currently suspended due to changes in funding available from Defra for bee health. Over the last 20 years, this work focused on *Varroa destructor* and the losses caused by honey bee virus infections that the mite transmits. In a collaborative project with Horticulture Research International (at University of Warwick), investigating potential biological control agents of *V. destructor*, the research identified and characterised fungal pathogens which are active against the mite but which are relatively safe for bees and other beneficial insects. Biological control offers an environmentally acceptable approach to the problem that could have considerable economic benefits, and we are actively seeking funding to continue this work.

We are currently analysing data from an Insect Pollinator Initiative funded project that assessed the impact of emergent diseases, including the Varroa associated Deformed wing virus, and the Microsporidian Nosema ceranae on the flight performance and orientation ability of honeybees and bumblebees and its consequences for bee populations.

HARMONIC RADAR
The use of harmonic radar in insect behaviour studies has been pioneered at Rothamsted. A transponder weighing just a few milligrams fitted to the thorax of bees picks up the interrogation radar signal and immediately emits a signal at a different frequency, which is then received by the radar. A recently awarded European Research Council grant will now enable cutting-edge development of the harmonic radar to allow us to collect data for entire adult life-spans and foraging ranges for multiple individuals of bee species, thus allowing us whole new insights into bee behaviour and pollination ecology.

INFORMATION EXCHANGE

Expertise in bee research is drawn upon by scientific colleagues world-wide and there are research links with institutes and universities in this country and abroad. Research findings are published in scientific journals but popular articles are also written for the beekeeping and agricultural press. Effective communication of our science by staff members is delivered via a vigorous programme of lectures presenting to national and local beekeeping associations and participation in various public media, including BBC programmes.

FUNDING

Rothamsted receives funds for research from the Biotechnology and Biological Sciences Research Council, through competitions and contracts from the Department for Environment, Food and Rural Affairs, the European Community, from Levy boards, commercial and other organisations. The support of the bee research programme in recent years by grants from the British Beekeepers Association, C. B. Dennis British Beekeepers Research Trust, the Eastern Association of Beekeepers and the Bedfordshire, Cambridgeshire, Norfolk, St Albans and Hertfordshire and High Wycombe Beekeepers Associations is gratefully acknowledged.

For more information visit: **http://www.rothamsted.ac.uk**

THE SCOTTISH BEEKEEPERS' ASSOCIATION

AIMS OF THE ASSOCIATION
- publish a monthly magazine
- maintain the Moir Library in Edinburgh
- conduct examinations in the art of beekeeping
- provide insurance and a compensation scheme for members

EDUCATION
The SBA arranges courses and awards certificates to successful candidates in the Scottish Basic Beemaster, Expert Beemaster, Honey Judge and Microscopy Examinations. It also actively promotes beekeeping by informing the public, especially the young, about bees and their benefits to the environment.

INSURANCE AND THE COMPENSATION SCHEME
All members of the SBA have insurance against Public Liability. The SBA Compensation Scheme is restricted to bee colonies located in Scotland and allocates part-replacement value for damage by vandalism, fire, theft and certain brood diseases.

LIBRARY
The SBA Moir Library in Edinburgh has one of the world's finest collections of beekeeping books. A library card is issued annually to every member who can borrow books at the cost of return postage only. Details may be obtained from the Library Convener.

MARKETS
Advice is given on all aspects of marketing honey products at appropriate times. Suggested bulk, wholesale and retail prices are notified in the magazine.

GENERAL SECRETARY
Tony Harris,
Cowiemuir,
Fochabers,
Moray
IV32 7PS
07884 496246
secretary@
scottishbeekeepers.org.uk

HON PRESIDENT
The Rt. Hon. Earl of Mansfield D.L, J.P
Scone Palace
Perth PH2 6BE

HON. VICE PRES,
Iain F Steven
4 Craigie View
Perth
PH2 0DP
01738 621100

Ian Craig
30 Burnside Ave, Brookfield,
Johnstone, Renfrewshire,
PA5 8UT
01505 322684
beekeeper30@btinternet.
com

HON. LIBRARIAN
Mrs. Margaret M. Sharp
City Librarian, City Library
George IV Bridge, Edinburgh

SBA

PUBLICATIONS
- The Scottish Beekeeper is published monthly and sent post free as part of the annual membership fee of £30 payable to the Membership Convener.
- Introduction to Bees and Beekeeping is £6.00 plus postage and may be obtained from the Advertising and Publicity Convener.

PUBLICITY
Members can purchase the Association tie, lapel badge, car sticker etc. Details may be obtained from the Advertising and Publicity Convener.

SHOWS
Two major annual honey shows are held in Scotland.

A honey competition and show with educational displays is held at the Royal Highland Show, Ingleston, Edinburgh in June and the Scottish National Honey Show is conducted at the Dundee Food and Flower Festival in September. Other Honey Shows are run in Ayr, Fife, Inverness, Turiff and at many other locations in Scotland as organised by Local Associations.

Executive Committee

CONVENERS OF STANDING COMMITTEES

MEMBERSHIP CONVENER
P. McAnespie
12 Monument Rd.Ayr
KA7 2RL   01292 885660
membership@
scottishbeekeepers.org.uk
INSURANCE & COMPENSATION
C. Irwin
55 Lindsaybeg Rd
Chryston, Glasgow
G69 9DW
0141 7791333
ceirwin@talktalk.net
ADVERTISING & PUBLICITY
Miss E Brown
Milton House, Main Street
Scotlandwell, Kinross
KY13 9JA   01592 840582
enidbrown6@gmail.com
EDUCATION,
Alan Riach
 Woodgate, 7 Newlands Ave,
Bathgate
EH48 1EE
01506 653839
alan.riach@which.net

PROMOTION OF BEEKEEPING
CONVENER

SHOWS,
Miss E Brown
Milton House, Main Street
Scotlandwell, Kinross
KY13 9JA
01592 840582
enidbrown6@gmail.com
LIBRARY,
Mrs Una Robertson
13 Wardie Ave
Edinburgh
EH5 2AB
una.robertson@btinternet.
com
MARKETS,
Margaret Thomas
Tig na Bruaich, Taybridge
Terrace, Aberfeldy,
Perthshire PH15 2BS
01887 829 710
zyzythomas@waitrose.com
BEE HEALTH,
Gavin Ramsay
Ealachan Bhana
Clachan Seil
Oban
PA34 4TL
01852 300383
ICT CONVENER,
Alasdair Joyce
Manachie Lodge.
Dallas Dhu
Forres
IV36 0RR
01309 671288
webmaster@
scottishbeekeepers.org.uk

AREA REPRESENTATIVES
NORTH,
Mrs Ann Chilcott
Sonas, Piperhill,
Nairn, Highland IV125SD
01667 404606
ann@chilcott.myzen.co.uk
EAST,
JOHN COYLE
Rose Cottage, Burnton,
By Kippen, Stirling
FK8 3JL
07774 266 540
 info@beekeepinginscotland.
co.uk
WEST,
Mike Thornley
Glenarn House, Glenarn
Road, Rhu, Helensburgh
G84 8LL
01436 820493
masthome@dsl.pipex.com
ABERDEEN AND MORAYSHIRE,
Dr Stephen Palmer
Fintry School House,
Fintry, near Turiff
AB53 5RN
01888 551367
palmers@fintry.plus.com

169

SBA

✉ ☎

S.B.A LECTURERS *Addresses in SBA Honey Judges List

All those listed may claim expenses except G. Sharpe, Scottish Bee Inspectors and Stephen Sunderland, all funded by SGRPID. All speakers accompany talks with visual aids

M Badger (General) *
0113 2945879
Miss. E. Brown (General) *
01592 840542

A. Chilcott
(Observation at the Hive
Entrance)
Sonas
Piperhill
Nairn
IV12 5SD
01667 404606
I. Craig (General) *
01505 322684

P.I Gibson
7 Shielswood Court
Galashiels
Selkirkshire
TD1 3RH
01896 750110
T. Harris* (Botany for
Beekeepers, Bumble and
Solitary Bees of Britain)
01343 821282

C. Irwin (General) *
0141 7791333

M.M. Peterson
(Bee genetics)
Balhaldie House,
High street, Dunblane
FK15 0ER
01786 822093

Dr G Ramsay
(Beekeeping on the Internet
/ Can Bees fight Varroa?)
Parkview, Station Road
Errol, Perth   PH2 7SN
01821 642385
A Riach * (Beehives through
the Ages, Microscopy))

Bryce Reynard, *Making
Skeps)
39 Old Mill Lane, Inverness
IV2 3XP
01463 225887

G. Sharpe (SAC) (Varroa
Management: My apiary
management system)
Apiculture Specialist
Life Science Technology
Group, SAC Auchincruive
Ayr KA6 5HW
01292 525375

Mrs M Thomas (General)
Tighnabraich, Taybridge
Terrace, Aberfeldy
Perthshire PH15 2BS
01887 829710

Scottish Bee Inspectors,
SGRPID, P Spur, Saughton
House,
Broomhouse, Edinburgh
EH11 3XD
0300 244 6672
beesmailbox@scotland.
gsi.gov.uk

Dr Peter Stromberg,
21 Woodside, Houston,
Renfrewshire,
PA6 7DD
01505 613 830
pstromberg1@aol.com

Stephen Sunderland,
Lead Bee Inspector,
SGRPID, P Spur, Saughton
House,
Broomhouse, Edinburgh
EH11 3XD
0300 244 6672
steve.sunderland@
scotland.gsi.gov.uk

Dr David Wright,
20 Lennox Row,
Edinburgh
EH5 3JW
0131 552 3439
bdwright20lr@btinternet.
com

MEMBER ASSOCIATIONS AND THEIR SECRETARIES

ABERDEEN,
Rosie Crighton
29 Marcus Cresc
Blackburn, Aberdeen
AB21 0SZ   01224 791181
aberdeenbeekeepers@
gmail.com

ARRAN BEE GROUP,
W K McNeish
Seafield, Kildonan,
Isle of Arran, KA27 8SE
01770 820357
wmcnsh@aol.com

AYR, Mrs L Baillie
Windyhill Cottage
Uplands Rd, Sundrum
Ayre, KA6 5JU
01292 570659
lbaillie@sundrum.demon.
co.uk

BORDER, Liz Howell
Oatlands, Houndridge,
Kelso
TD5 7QN
01573 470747
kevhwl@aol.com

BUTE, Alison Cross
Marionslea,
Minister's Bray, Rothsay,
Isle of Bute
PA20 9BG
01700 504627
alison.cross2@virgin.net

CADDONFOOT,
James & Julia Edey
West Water, Bedrule,
Hawick, Roxburghshire,
TD9 8TD
01450 870400
jamesedey@googlemail.
com

COVINGTON AND THANKERTON,
Angus Milner-Brown
Covington House,
Covington Road, Biggar
ML12 6NE
01899 308024
angus@therathouse.com

COWAL, Brian Madden
123a Alexandra Parade
Dunoon, PA23 8AW
01369 703317
brian_maden@btinternet.
com

DINGWALL, Alpin Stewart
Rowan Cottage, Fasaig,
Torridon by Achnasheen,
Ross-shire
IV22 2EZ
01445 791450
dingwall.beekeeping@
googlemail.com

DUNBLANE & STIRLING,
Fiona Fernie
Greystones Dunira,
By Comtie
PH6 2JZ
01764 679152
secretary@
dunblanebeekeepers.com

DUNFERMLINE & WEST FIFE
Dr T Scott
Grange Farmhouse
Grange Rd, Dunfermline
KY11 3DG
01383 733125
dwf@fifebeekeepers.couk

EAST LOTHIAN,
Deborah Mackay
5 Goshen Farm Steading,
Musselburgh, East Lothian
EH21 8JL
0131 665 8939
eastlothianbeekeepers@
googemail.com

EAST OF SCOTLAND,
Colin Smith
The Laundry House,
Ethie, Inverkielor, Arbroath,
DD1 5SP
secretary@
eastofscotlandbeekeepers.
org.uk

EASTER ROSS,
Colin Ridley
Stirling Cottage,
Lamington Park, Kildray,
Ross-shire
IV18 0PE
01862 842410
colinr031@googlemail.com

EASTWOOD,
Robert Gordon
10 Caribar Drive,
Barrhead, Glasgow
G78 1BQ
0141 5716498
Robert.gordon@ntworld.
com

EDINBURGH & MIDLOTHIAN
Gordon Jardine
20 Pentland Grove,
Edinburgh,
EH10 6NR
07703 528801
gordieric@hotmail.com

FIFE,
Janice Furness
The Dirdale, Boarhills
St. Andrews, Fife KY16 8PP
01334 880 469
jcfurness@dirdale.fsnet.
co.uk

FORTINGALL,
Mrs. Jo Pendleton,
Lilac Cottage
Old Bridge of Tilt by
Pitlochry
PH18 5TP
01796 481 362
d.h.pendleton@btinternet.
com

GLASGOW DISTRICT,
Mhairi Neill
3 Machan Ave, Larkhall,
ML9 2HE
01698 881602
glasgowbeeksec@hotmail.
co.uk

HELENSBURGH,
Gordon Smith
The Moorings,
Ferry Road, Rhu,
G84 9PY
07980 578206
Gordon@windsmiths.co.uk

HONEYPOTZ (WEST LOTHIAN),
Dave Gillan
34 Gardner Crescent,
Whitburn,
EH47 0PE
01501 744817
honeypotzbeekeeping@live.
co.uk

INVERNESS,
Julia Moran
Woodend Cottage,
Dunloit, Drumnadrochit,
IV36 6XF
01456 450463
woodendcottage@homecall.
couk

KELVIN VALLEY,
I Ferguson
4 South Glassford Street
Milngavie
G62 6AT
0141 956 3963
jeanian@ferguson2007.plus.
com

KILBARCHAN AND DISTRICT,
I. Craig
30 Burnside Ave
Brookfield
Johnstone PA5 8UT
01505 322684
beekeeper30@btinternet.
com

KILMARNOCK & DISTRICT,
J. Campbell
North Kilbryde House,
Stewarton, Kilmarnock,
KA3 3EP
01560 482489
john.d.campbell@talktalk.net

KINTYRE & MID ARGYLL,
Zoe Weir
1 Lagnagorton,
Clachan, Tarbert,
Argyll, PA29 6XW
lagnagortan@aol.com

LARGS & DIST,
Ruth Anderson
07773 776253
Ruthanderson7@
googlemail.com

LOCHABER,
Sarah Kennedy
Tigh na Feid,
Achintore Road,
Fort William
PH33 6RN
secretary@
lochaberbeekeepers.org

MORAY,
Anne Black
Four winds,
Prospect Terrace,
Lossiemouth, Moray
IV31 6JS
01343 810899
secretary@
moraybeekeepers.co.ukcom

MULL,
Mrs. S. Barnard
Viewmount, Tobermory
Isle of Mull PA75 6PG
01688 302008
tim-barnard@lineone.net

NAIRN & DISTRICT,
Ruth Burkhill
2 Cloves Cottage,
Alves, Moray
IV36 2RA
01343 850041
ruthburkhill@gmail.com

NEWBATTLE,
Joyce Jack,
23 South Park West,
Peebles EH45 9EF
01721 722444
joycecjack@aol.com

OBAN & DISTRICT,
Phil Moss
An Isean Eala
Clachan Seil
Oban PA34 4TL
01852 300383
aniseaneala@btinternet.com

OLRIG AND DISTRICT,
Robin Inglis
Roadside Skirza
Freswick, Wick KW1 4XX
01955 611260
gailinglis@btinternet.com
ORKNEY,
Sue Spence
Alton House,
Berstane Road,
Kikwall, Orkney
KW15 1NA
01856 873920
bs3920@yahoo.com
PEEBLES-SHIRE,
Amanda Clydesdale
20 Kingsmeadows Gardens
Peebles EH45 9LB
01721 720563
amanda.clydesdale@
btinternet.com
PERTH AND DISTRICT
Brian Clelland
12 Albert Road,
Scone, Perth,
PH2 6QH
07845375298
info@
perthanddistrictbeekeepers.
co.uk
SKYE & LOCHALSH,
Joe Grimson
1 Riverside Cottage,
Braeintra, Stromeferry,
IV53 8UP
j.grimson@btinternet.com

SOUTH OF SCOTLAND,
Debbie Park
Crofthead,
Dalswinton,
Dumfries
DG2 0XY
01387740030
d.park@yahoo.co.uk
SPEYSIDE,
Gerry Thompson
Highland House,
Knockando,
Aberlour, Moray,
AB53 7RP
01340810229
gjthom@sky.com
SUNART, ARDNAMURCHAN,
MOIDART AND MORVERN
Kate Atchley
Anasmara,
Mingarry, Acharacle,
Argyll & Bute
PH36 4XJ
07774807645
bees@kateatchley.co.uk
SUTHERLAND,
Sue Steven
Mulberry Croft,
2 East Newport,
Berriedale Caithness
KW7 6HA
01539 751 245

SBA

SBA ACTIVE HONEY JUDGES

M BADGER
Kara, 14 Thorn Lane,
Roundhay,
Leeds
LS8 1NN

MISS E. BROWN
Milton House, Main Street,
Scotlandwell, Kinross
KY13 9JA
01592 840582
01397 712730

M. CANHAM
Whinhill Farm House
 Nairn IV12 5RF
01667 404314

I. CRAIG
30 Burnside Avenue
Brookfield,
Johnstone Renfrewshire,
PA5 8UT
01505 322684

H DONOHOE
7 Grant Road
Banchory
AB31 5UW
01330 823502

T. HARRIS
Cowiemuir
Fochabers,
Moray IV32 7PS
01343 821282

C. E. IRWIN
55 Lindsaybeg Road
Chryston, Glasgow
G69 9DW
0141 7791333

DR F. ISLES
"Gardenhurst",
Newbigging
Broughty Ferry
Dundee DD5 3RH
01382 370315

P MATHEWS
MRS C MATHEWS
4 Annanhill
Annan, Dumfries-shire
DG12 6TN
01461 205525

MS B L MCLEAN
Upper Flat, 2 Invererne Rd,
Forres IV36 1DZ
01309 676316

A. RIACH
7 Newlands Avenue,
Bathgate
EH48 1EE
01506 653389

C. WEIGHTMAN
Shilford, Stocksfield,
Northumberland
NE43 4HW
01661 842082

C. WILSON
Cedarhill, Auchencloch,
Banknock, Bonnybridge
FK4 1VA
01324 840227

DR D WRIGHT
MRS B WRIGHT
20 Lennox Row
Edinburgh
EH3 5JW
0131 552 3439

M. YOUNG
101 Carnreagh,
Hillsborough
County Down
N. Ireland
BT26 6LJ
0289 268972

ULSTER BEEKEEPERS' ASSOCIATION

www.ubka.org

OBJECTS OF THE ASSOCIATION
The objects of the Association are to unite beekeepers for their mutual benefit to serve the best interests of beekeeping by all means within its power and to foster its healthy development.
For the purpose of achieving these objects the Association will:
- promote the formation of local Beekeepers' Associations
- disseminate information and advice about beekeeping
- provide examination facilities in the craft of beekeeping
- encourage maintenance and improvement of the beekeeping environment.

EDUCATION
In conjunction with the College of Agriculture, Food & Enterprise (CAFRE), the U.B.K.A. assists in organising classes for Preliminary, Intermediate and Senior Certificate Examinations in Beekeeping following the syllabus of the Federation of Irish Beekeepers' Associations (FIBKA).

INSURANCE
Affiliated local Associations and their individual members have access to the UBKA group public and product liability insurance scheme.

APIARY SITES
Almost all twelve local Associations and CAFRE's Greenmount Campus have access to apiary sites and, for some sites, access to observation houses provided with help from Leader 2 funding, for use in demonstrating and promoting good practice to members, schools and other interested groups.

PRESIDENT,
David Wright
24 Quarry Road
Lisbane, Comber,
Newtownards
Co Down BT23 5NF

SECRETARY,
Lorraine McBride,
11 Ballyloughan Park,
Ballymena,
BT43 5HW.
email:
ubkasecretary@gmail.com

TREASURER,
Gail Orr.
64 Ballycrune Rd.,
Hillsborough,
BT26 6NH

LECTURERS
Vanessa Drew,
40 Lacken Rd.,
Ballyroney,
Banbridge,
Down BT32 5JA

Jim Fletcher
26 Coach Road, Comber
Co.Down. BT23 5QX

UBKA

✉ ☎

Lecturers continued
Ethel Irvine
2 Laragh Lee
Ballycassidy
ENNISKILLEN
BT94 2JT

Lorraine McBride
11, Ballyloughan Park
Ballymena, Co.Antrim,
BT43 5HW

Rev Sam Millar
41 Rectory Park
Garvagh, COLERAINE
Co Londonderry
BT51 5AJ

Norman Walsh
43, Edentrillick Rd
Hillsborough, Co. Down
BT26 6PG

HONEY SHOWS

Local Associations stage honey shows throughout Northern Ireland. The Northern Ireland Honey Show, hosted by the Belfast City Parks Department, is held annually in September in the Botanic Gardens Belfast.

CONFERENCE

The 71st UBKA Annual Conference will be held on 20th – 21st March 2015 at CAFRE's Greenmount Campus, Antrim. Contact the U.B.K.A. Conference Secretary at 07871 161303 and www.ubka.org for details.

SECRETARIES OF ASSOCIATIONS

Belfast,
Jonathan Getty
80 Locksley Park,
Belfast,
BT10 0AS
email: bbkasecretary@
googlemail.com

Clogher Valley,
Chester Roulston
10 Ednagee Rd,
Garvetagh, Castlederg,
Co. Tyrone
BT81 7QF
email: chester.roulston@
hotmail.co.uk.

Derry City, Jen Simpson
4 Cullinean Manor,
Redcastle,
Nr Lifford
Co. Donegal
email:
jennifersimpson163@
yahoo.com.

Dromore, Patrick Lundy
116 Dromore Road,
Ballynahinch,
Co.Down
BT24 8HK
email: patrickjlundy@
gmail.com

East Antrim, Stephen
Robinson
53 Wellington Ave.,
Larne, Co Antrim
BT40 1EH
email: admin@sgrobinson.

co.uk.

Fermanagh,
Jorgen Pedderson
email: enniskillencc@
gmail.com.

Killinchy, Dawn Stocking
Ballycruttle House,
7 Tullynaskeagh Road
Downpatrick
Co. Down
BT30 7EJ
email: kbkasecretary@
gmail.com

Mid Antrim, Angela Morrow
23 Beechwood Drive
Ahogill
Ballymena
BT42 1NB
email: midantrimbka@
btinternet.com

Mid Ulster, Anne Milligan
61 Blackisland Road ,
Annaghmore,
Portadown
BT62 1NE
email: annpmilligan@
gmail.com

Randalstown, Brian Gillanders
email: hightown@live.co.uk

Roe Valley, Sandra Logan
22 Knocknougher Road,
Macosquin,
Coleraine,
Co. Londonderry.

email: alexandramlogan@
aol.com
The Three Rivers,
Daniel McMenamin
email: daniel@urney.info.

Rostrevor & Warrenpoint,
Darren Nugent,
email: biglongdarren@
hotmail.com

Honey Judges
Jim Fletcher
26 Coach Road,
Comber
BT23 5QX

Michael Young
Mileaway, Carnreagh
Road
Hillsborough
Co. Down
BT26 6LJ

Norman Walsh
43 Edentrillick Rd
Hillsborough
Co. Down
BT26 6NH

CYMDEITHAS GWENYNWYR
CYMRU WELSH BEEKEEPERS' ASSOCIATION

AMCANION Y GYMDEITHAS / AIMS OF THE ASSOCIATION
- Promote and develop beekeeping in Wales
- Conduct examinations in beekeeping
- Liaise with organisations and bodies for the benefit of beekeeping in Wales

AELODAETH UNIGOL / INDIVIDUAL MEMBERSHIP
Individual membership of the WBKA is provided for persons who do not live within the areas of branch associations, and wish to support the association. Information relating to benefits and facilities provided for individual members is available from the Individual Membership Secretary.

ARHOLIADAU / EXAMINATIONS
The Examinations Board conducts six grades of examinations: Junior, Primary, Intermediate, Practical, Honey Show Judges, Senior. Information is available from the Examination Board Secretary.

Candidates following the Duke of Edinburgh Award Scheme may receive information regarding the inclusion of beekeeping as a course submission from the Examinations Secretary.

CYNHADLEDD/ CONVENTION
At the Royal Welsh Agricultural Society's Showground, Llanelwedd. This event is normally held during Late March/ Early April. Information relating to this event is available from the convention secretary.

YSWIRIANT / INSURANCE
All individual and fully paid up members of beekeeping associations affiliated to WBKA are covered against 'Public and Product' liability claims. All affiliated associations are covered against public liability during conventions officially organised by the association.

YSGRIFENNYDD / SECRETARY
John Page
The Old Tannery
Pontsian
Llandysul
Ceredigion
SA44 4UD
secretary@WBKA.com

LLYWYDD/PRESIDENT
David Culshaw,
9 Ash Grove, Llay,
Wrexham LL12 0UF
president@wbka.com

CADEIRYDD/CHAIR
Jenny Shaw
Llwyn Ysgaw,
Dwyran, Llanfairpwll,
Anglesey LL61 6RH
chair@wbka.com

IS-GADAIRYDD/
VICE CHAIR
Sue Townsend.
depchair@wbka.com

WBKA/CGC

TRYSORYDD/TREASURER
Margaret Jones

GWEFEISTR/WEBMASTER
Grant Williams

AND GOLYGYDD/EDITOR
Sue Closs
editor@wbka.com

IS-OLYGYDD (ERTHYGLAU CYMRAEG)/SUB EDITOR
Dewi Morris Jones
Llwynderw, Bronant
Aberystwyth SY23 4TG
(01974 251264)

ARHOLIADAU/EXAMINATIONS
Lynfa Davies

The WBKA Individual Membership benefits include cover under the BDI Scheme against the loss, due to foul brood diseases, of a minimum number of stocks (determined by BDI). Affiliated Associations provide this cover for their members.

LLYFRGELL / LIBRARY
The reference sections of all county libraries in Wales have details of the names and addresses of Secretaries of Associations affiliated to WBKA.

Books on beekeeping can be borrowed from county, branch and mobile libraries. The Library, Ffordd y Bala, Dolgellau LL40 2YS, has been nominated to stock beekeeping books.

Members of associations affiliated to IBRA may borrow books/documents from its library.

GWENYNWYR CYMRU - The Welsh Beekeeper
A publication of the Welsh Beekeepers Association, giving news and views of beekeeping and related subjects. Articles and advertisements enquiries should be sent to the Editor. Articles written in Welsh should be sent to the Sub Editor. Gwenynwyr Cymru is provided free to members of Affiliated Associations and Individual Members. Information regarding subscriptions is available from the Individual Membership / Subscription Secretary.

DARLITHWYR / DANGOSWYR, LECTURERS / DEMONSTRATORS
The names and addresses of lecturers and demonstrators, recommended by associations affiliated to the WBKA, are available from the General Secretary.

CYNLLUN CYSWLLT CHWYSTRELLU / SPRAY LIAISON SCHEME
Information is available from the General secretary

SIOEAU / SHOWS

Honey/beekeeping sections are included at the Royal Welsh Agricultural Show, Llanelwedd, (OS ref: SO040520) during July, and at county, town and village shows throughout Wales. Information relating to these events may be obtained from secretaries of associations in the locality of the shows.

The historic FFAIR FEL ABERCONWY is held annually in the main street of the town, (OS ref: SH278378), on 13th September. Further information is available from the secretary of Conwy Association.

RHEOLAU CYFREITHIOL / STATUTORY REGULATIONS

The administration of the statutory regulations governing all aspects of beekeeping in Wales, is the responsibility of the Wales National Assembly, Caerdydd, CF99 1NA Phone (02920) 825111 Fax: (02920) 823352 Matters concerning statutory regulations, their implications and execution, should be addressed to the Minister of Agriculture and Rural Affairs, Wales National Assembly, at the above address.

AELODAETH UNIGOL- TANYSGRIFAU/ INDIVIDUAL MEMBERSHIP SUBSCRIPTIONS

Ian Hubbuck
White Cottage
Berriew
SY21 8BB
01686 640205
ianhubbuck@hotmail.com

INSURANCE:

Sue Townsend
insurance@wbka.com

CONVENTION SECRETARY:

Graham Wheeler
mertyndowning@
btinternet.com

CONVENTION TRADE STANDS SECRETARY:

Wally Shaw
Llwyn Ysgaw, Dwyran,
Llanfairpwll, Anglesey
LL61 6RH 01248 430811
waltershaw301@
btinternet.com

WBKA/CGC

CYMDEITHASAU TADOGOL A'U YSGRIFENYDDION /
AFFILIATED ASSOCIATIONS AND SECRETARIES

ABERYSTWYTH, Ann Ovens,
Tan-y-Cae, Nr Talybont,
Ceredigion,
SY24 5DP
01970 832359
ann.ovens@btinternet.com
ANGLESEY,Jim Allen,
Pwll yr Olwyn, Dulas,
Ynys Mon LL70 9EX,
01248 410338
secretaryabka@gmail.com
BRECKNOCK AND RADNOR,
Dr Gillian Todd, Meadow
Breeze, Llanddew, Brecon
LD3 9ST
01874610902 07971314798
gbtodd@btinternet.com
BRIDGEND, Sue Verran
Ty Mel, Maesteg Rd.
Bridgend
CF32 0EE
01656 729699
verran@btinternet.com
CARDIFF AND VALE, Annie
Newsam
Stonecroft, Mountain Road,
Bedwas, Caerphilly, CF83
8ER
annienewsam@hotmail.co.uk
CARMARTHEN, Brian Jones
Cwmburry Honey Farm,
Ferryside, Carmarthenshire,
SA17 5TW
01267 267318
beegeejay2003@yahoo.co.uk

CONWY, **Mr Peter McFadden,**
Ynys Goch
Ty'n y Groes,
Conwy LL32 8UH
01492 650851
peter@honeyfair.freeserve.
co.uk
EAST CARMARTHEN
Geoff Saunders,
Pant yr Esgair, Carmel Road,
Talley, Llandeilo SA19 7DX,
01558 685331.
secretary@ec-bka.com
FLINT AND DISTRICT,
Jill and Graham Wheeler,
Mertyn Downing, Whitford
Holywell, Flintshire,
CH8 9EP.
01745 560557
mertyndowning@btinternet.
com
GWENYNWYR CYMRAEG
CEREDIGION W.I.Griffiths,
Llain Deg, Comins Coch,
Aberystwyth, SY23 3BG
01970 623334
 wilmair@btinternet.com
LAMPETER AND DISTRICT
Mr Gordon Lumby,
Gwynfryn, Brynteg,
Llanybydder,
SA40 9UX
01570 480571
g.lumby@btopenworld.com

LLEYN AC EIFIONYDD
Amanda Bristow,
Bryngwydion, Pontllyfni,
Gwynedd
LL54 5EY 01286 831328
amanda.bristow@egnitec.com
MEIRIONNYDD,
Sue Townsend,
01341 430262
suetownsend@tesco.net
MONTGOMERYSHIRE,
Keith Rimmer,
Beudy Clyd, Maesmawr,
Caersws SW17 5SB,
01686 689061
secretary@montybees.org.uk

PEMBROKESHIRE,
John Dudman
01437 891892
secretarypbka@hotmail.com
SOUTH CLWYD,
Mrs Carol Keys-Shaw, Y Beudy,
Maesmor Hall, Maerdy
Corwen LL21 0NS
01490 460592
c.keysshaw@btinternet.com
SWANSEA, Paul Lyons,
2 West Cliff, Southgate,
Swansea, SA3 2AN.
paul.lyons@bt.com

HEB DADOGU/NON
AFFILIATED:
Mrs J Bromley
Ty Hir, Monmouth Road
Raglan, Usk. NP15 2ET
01291 690331
bromleyjan@hotmail.com

BEIRNIAID SIOE FÊL TRWYDDEDIG / WBKA QUALIFIED HONEY SHOW

TERRY E. ASHLEY
Meadow Cottage,
11 Elton Lane, Winterley
Sandbach CW11 4TN
M. J. BADGER MBE
14 Thorn Lane, Leeds
LS8 1NN
M BESSANT
Gwili Lodge, Heol
Lotwen, Rhydaman
SA18 3RP
ROBERT BREWER
PO Box 369, Hiawassee,
Georgia, USA
TOM CANNING
151 Portadown Road,
Armagh, Co Armagh
BT61 9HL
LES CHIRNSIDE
Bryn-y-Pant Cottage,
Upper Llanover,
Abergavenny NP7 9ES

CARYS EDWARDS
Ty Cerrig, Ganllwyd,
Dolgellau LL40 2TN
IFOR C. EDWARDS
Lleifior, Pontrhydygroes,
Ystrad Meurig SY25 6DN
STEVEN GUEST
Bridge House, Hind
Heath Road, Sandbach,
CW11 3LY
HUGH MCBRIDE
11 Ballyloughan Park
Antrim BT43 5HW
LORRAINE MCBRIDE
11 Ballyloughan Park
Antrim BT43 5HW

CECIL MCMULLAN
33 Glebe Road,
Hillsborough, County
Down
LEO MCGUINESS
89 Dunlade Road, Grey
Steel BT47 4QL
GAIL ORR
64 Ballycrone Road,
Hillsborough BT26 6NH
DINAH SWEET
Graig Fawr Lodge,
Caerphilly, CF83 1NF
REDMOND WILLIAMS
Tincurry, Cahir, Co
Tipperary Eire
MICHAEL YOUNG MBE
Mileaway, Carnreagh,
Hillsborough BT26 6LJ

NBU

✉ ☎

NATIONAL BEE UNIT, Animal and Plant Health Agency (APHA) – Formally part of The Food and Environment Research Agency (FERA)

National Bee Unit
APHA, National Agri-Food
Innovation Campus
Sand Hutton, York, YO41
1LZ, UK

Tel.No: 03003030094
Fax.No: 01904 462240
E-Mail: nbu@fera.gsi.gov.uk
Website:
www.nationalbeeunit.com

Earlier this year it was announced that a combined agency would be created, with four functions of Fera (Bee inspectorate, the Plants Health and Seeds Inspectorate, the Plant Variety and Seeds Group and the GM Inspectorate) joining with AHVLA. From 1 October we will become part of the new Animal and Plant Health Agency. Animal and Plant Health Inspectors have a strong history of working together in times of disease emergency, and this will be made easier when they are part of the same organisation. The Animal and Plant Health Agency will also play a vital role in stopping pests, diseases, and invasive non-native species entering the UK.

www.nationalbeeunit.com

NATIONAL BEE UNIT TECHNICAL
STAFF, HEAD OF UNIT
Mike Brown

NATIONAL BEE INSPECTOR
Andy Wattam
01522 789726
07775 027524

National Bee Unit

The National Bee Unit (NBU) is part of APHA, an agency of the Department for Environment, Food and Rural Affairs (Defra), and is based just outside York. The NBU is an element of APHA and its work covers all aspects of bee health and husbandry in England and Wales, on behalf of Defra in England and for the Welsh Government in Wales. The work of the unit includes disease and pest diagnosis, research into bee health matters, development of contingency plans for emerging threats, import risk analysis, related extension work and consultancy services to both government and industry.

Bee Health Inspection Service

The Integrated Bee Health Programme is run by the NBU on behalf of core policy customers. The NBU has a long track record in bee husbandry and bee disease control (since 1946) and has been directly responsible for the Bee Inspection services in England and Wales since 1994. The NBU consists of a home-based

Inspectorate team, and the laboratory diagnostic and research team based at National Agri-Food Innovation Campus, Sand Hutton. In addition colleagues across Fera contribute to the programme and research projects. The Bee Health Inspectorate team consists of approximately 60 home-based members of staff. It is headed by the National Bee Inspector (NBI), whose role it is to manage the statutory disease control and training programmes. The NBI has management responsibility for eight home-based Regional Bee Inspectors (RBIs), one heading each of the seven regions in England and one covering Wales. The RBI in turn manages a number of Seasonal Bee Inspectors (SBIs). The RBIs and SBIs organise inspections under EU and UK legislation, submit suspect samples for diagnosis, treat colonies for foulbrood and train beekeepers in bee husbandry for better disease control and greater self-sufficiency. In addition the Bee Inspectors also collect honey samples for residue analysis under the Statutory Honey collection agreement with Defra Veterinary Medicines Directorate (VMD). With Aethina tumida (Small hive beetle (SHB)) and Tropilaelaps spp. both notifiable under UK and EU law, Inspectors also undertake surveillance for these exotics in "Sentinel Apiaries (SA)" close to identified high risk areas. Beekeepers who manage SA's represent a valuable front line defence against an exotic pest incursion.

Bee Disease Diagnostic Team
The NBU's diagnostic team provides a rapid, modern service for both the Inspection team and beekeepers. The NBU laboratory adheres to Good Laboratory Practice (GLP) codes which is a quality accreditation scheme administered by the Department of Health. All diagnostic tests are conducted according to the OIE (Office International des Epizooties) Manual of Standard Diagnostic Tests and Vaccines. The OIE is the World Organisation for Animal Health and produce internationally recognised disease diagnosis guidelines (http//www.oie.int.) Across Fera diagnostic support is provided from teams of microbiologists acarologists, insect virologists and molecular specialists.

REGIONAL BEE INSPECTORS
Ian Molyneux
Northern Region
01204 381186
07775 119442

Charles Millar
Western Region
01694 722419
07775 119476

Nigel Semmence
Southern Region
01264 338694
07776 493649

Julian Parker
South East Region
01494 578505
07775 119469

Simon Jones
South West Region
01823 442228
07775 119459

Keith Morgan
Eastern Region
01485 520838
07919 004215

Ivor Flatman
North East Region
01924 252795
07775 119436

Frank Gellatly
Wales
01558 650663
07775 119480

To find details of Seasonal /Bee Inspectors please see BeeBase: https://secure.fera.defra.gov.uk/beebase/index.cfm . Remember that Seasonal Inspectors only work from April to September

Research Co-ordinator
Kat Roberts

Research Scientist
Claire Webster

Science Coordinator
Gay Marris

Laboratory
Ben Jones (laboratory manager)
Ruth Grant
Ilex Whiting
Victoria Tomkies

Apiarists
Damian Cierniak
Jack Wilford

Technical Advisor
Jason Learner

**Administrative
Programme Support**
Kate Parker,
Lesley Debenham
& Jenna Cook

Bees and the Law

The 1980 Bees Act empowers Ministers to make Orders to control pests and diseases affecting bees, and provides powers of entry for Authorised Persons. Under the Bees Act, The Bee Diseases and Pests Control Order 2006 for England and Wales, (there is similar legislation for Scotland and Northern Ireland) designates American foulbrood (AFB), European foulbrood (EFB), *A. tumida* (SHB) and *Tropilaelaps* mites (all species) as notifiable pests and defines the action which may be taken in the event of outbreaks. The Trade in Animals and Related Products Regulations 2011 (TARP) give effect to EU legislation concerning trade in animals and animal products from other Member States and the importation of animals and animal products from Third Countries. Consolidated in TARP is the Directive on animal health requirements for trade in bees (the Balai Directive (92/65/EEC)). The Balai Directive specifically describes the intra-trade certification requirements for honey bees. Annexes A and B list the pests and diseases of animals, including those that affect honey bees, which are considered highest risk. Annex A lists the notifiable organisms throughout the Union. These are American foulbrood (AFB), the Small hive beetle (*A. tumida*) and *Tropilaelaps* mites. Annex B lists organisms that are not notifiable across the EU, but which Member States may choose to cover under their own domestic legislation. At the time of writing time neither the small hive beetle nor *Tropilaelaps* have been confirmed in Europe.

The Importation of Bees

Beekeepers may legally import Queen honey bees from listed Third Countries as set out in the Commission Decision 2005/60/EC. The list of countries is currently restricted to, Argentina, Australia and New Zealand. Honey bee queens and packages may be imported from the EU, however, the beekeeper must make themselves aware of and adhere to the importation guidelines which can be found on BeeBase:

https://secure.fera.defra.gov.uk/beebase/index.cfm?
pageid=126.
Under the Balai directive consignments of bees moved between Member States must be accompanied by an original health certificate confirming freedom from notifiable pests and diseases. In addition, colonies may be subject to controls aimed at preventing the spread of Fireblight between 15th March and 30th June. More information about this can be found on BeeBase https://secure.fera.defra.gov.uk/beebase/index.cfm?pageid=103

American and European foulbrood
American and European foulbrood are both serious diseases of European honey bees and are subject to statutory control in the United Kingdom.
Any beekeeper who suspects their colonies to be infected with foulbrood should contact their nearest Appointed Bee Inspector (ABI) to report this.
Those apiaries that are suspected or confirmed to have a notifiable disease will be issued with a Standstill notice prohibiting the movement of any hive, bees, combs, bee products, bee pests, hive debris, appliances or other things liable to spread a suspected notifiable disease or pest on those premises or vehicle except under licence.
More information about foulbrood disease is available on our website or in our advisory leaflet 'Foulbrood Disease of Honey Bees and other common brood disorders' https://secure.fera.defra.gov.uk/beebase/index.cfm.

Varroa
As part of the NBU's routine field screening programme the first known case of pyrethroid resistant Varroa mites in the UK was discovered in apiaries in Devon in August 2001. The NBU undertook a resistance-monitoring programme throughout England and Wales. Pyrethroid resistant *Varroa* mites are now widespread in England and Wales. To access current advice on *Varroa* and it's management please visit BeeBase.

Adult Bee Diseases
The NBU also offers a non-statutory and chargeable service for screening adult bee diseases. The NBU tests

187

for adult bee diseases and parasites such as Nosema species (Nosema apis and *Nosema ceranae*, amoeba (*Malpighamoeba mellificae*) and tracheal mites (Acarine or *Acarapis woodi*) from samples submitted by beekeepers. As these diseases are non-statutory this service is chargeable. For the current cost please contact the NBU or visit the website.

Exotics
Beekeepers must make themselves aware of the potential threats to beekeeping in the UK. The field Inspection team monitors for potential exotics, the SHB and *Tropilaelaps spp*. The laboratory team also routinely screen import samples and suspect samples submitted for identification by both beekeepers and the field team.

Pesticide Monitoring
The Wildlife Incident Investigation Scheme (WIIS) is a unique scheme for monitoring the effects of pesticides on wildlife, including beneficial invertebrates such as honey bees. It is led by the Chemicals Regulation Directorate (CRD) with Natural England managing and undertaking site enquiries on their behalf; The Food and Environment Research Agency (Fera) carry out disease and pesticide analysis and, if appropriate, the Veterinary Laboratories Agency (VLA) carry out post mortems on wildlife. Information gathered is fed into the approval process for pesticides and helps in the verification and improvement of pesticide risk assessments.
It can also result in changes to label recommendations on pesticide products. It is not provided as a personal service to beekeepers wishing to seek evidence for the purpose of civil litigation but can lead to enforcement action being taken by the enforcer if the misuse or abuse of a product is identified as part of this process. For more information please see BeeBase: https://secure.fera.defra.gov.uk/beebase/index.cfm?pageId=84.

Research & Development
A programme of research and development within the group underpins the Unit's work. They also have long-established links with many European and world wide research centres, universities and the beekeeping industry. The primary aim of our R&D is to improve our understanding of the issues which impact bee health. The NBU also actively supports PhD students, some of which are funded using donations from the beekeeping industry.
For an update on the current R&D work of the unit please see BeeBase.

Extension

The NBU trains beekeepers in several ways: local courses and advisory visits run by the Inspectors, and national courses held at the York laboratory. The NBU annually hosts the National Diploma in Beekeeping residential courses and has also been host to visiting overseas workers and researchers. NBU York based staff also provide training to beekeepers at local and regional beekeeper meetings.

Healthy Bees Plan

The Healthy Bees Plan was published by Defra and the Welsh Assembly Government in March 2009 following consultation with beekeepers and the main Beekeeping Associations. It sets out a plan for Government, beekeepers and other stakeholders to work together to respond effectively to pest and disease threats and to put in place programmes to ensure a sustainable and productive future for beekeeping In England and Wales. The Healthy Bees Plan consists of three working groups that report to the project management board to help deliver the five major objectives of the plan. To view the Healthy Bees Plan, please see BeeBase.

BeeBase

BeeBase is the National Bee Unit website. It is designed for beekeepers and supports Defra, WAG and Scotland's Bee Health Programmes and the Healthy Bees Plan, which set out to protect and sustain our valuable national bee stocks. Our website provides a wide range of free information for beekeepers, to help keep their honey bees healthy. We hope both new and experienced beekeepers will find this an extremely useful resource and sign up to BeeBase. Knowing the distribution of beekeepers and their apiaries across the country helps us to effectively monitor and control the spread of serious honey bee pests and diseases, as well as provide up-to-date information in keeping bees healthy and productive. By telling us who you are you'll be playing a very important part in helping to maintain and sustain honey bees for the future. To register as a beekeeper please visit BeeBase.

DARD

✉ ☎

DEPARTMENT OF AGRICULTURE AND RURAL DEVELOPMENT

WWW.dardni.gov.uk

BEE DISEASE DIAGNOSTICS:
Sam Clawson
Agri-Food and Biosciences
Institute (AFBI)
Newforge Lane
BELFAST BT9 5PX
Tel: 028 9025 5289
Email:
sam.clawson@afbini.gov.uk

Training Courses:
Hazel Marcus
Greenmount Campus
College of Agriculture Food
and Rural Enterprise
Tel: 028 9442 6964
Email:
industry.trainingadmin@dardni.
gov.uk

Bee Health Inspections:
Thomas Williamson
Plant Health Inspection Branch,
DARD, Glenree House,
Carnbane Industrial Estate,
Newry, Co Down, BT35 6EF
Tel: 028 3889 2374
Fax: 028 3025 3255
Email:
tom.williamson@dardni.gov.uk

Honeybee Regional Report for Northern Ireland 2014

Bee Health Surveys

A questionnaire survey for Bee Husbandry issues has been circulated annually to beekeepers via beekeeping associations since 2009. The results of the 2013 survey are available as a pdf on the AFBI website (www.afbini.gov.uk). Preliminary results from the 2014 survey show colony loses were 8% compared to 43% in 2013. 77 percent of responding beekeepers reported no losses. The 2014 survey results are currently being processed but will be available on the AFBI website in November.

Bee Health Inspections

The Bee Inspectorate carried out surveys for American foul brood, European foul brood, Small Hive beetle and Tropilaelaps mite. American foul brood remains a problem for beekeepers in Northern Ireland with 7 apiaries found to have the disease by early September compared to 14 apiaries in total for 2013. Inspections were also carried out for European foul brood without any outbreaks recorded. Surveys continued for Small Hive Beetle and Tropilaelaps mite. Apiaries in the vicinity of ports or fruit importers were targeted for Small Hive Beetle inspections using corriboard shelter traps, while apiaries that had imported in the past were selected and hive scrapings examined for Tropilaelaps mite.

Adult Bee Disease Diagnostics

Nosema ceranae was first recorded in Northern Ireland in 2010. *N. ceranae* is an emergent pathogen of western honeybees. It is similar to the endemic species, *Nosema apis* but is considered to produce a more virulent disease than *N. apis*, probably reflecting its more recent association with the western honeybee. In 2013, 201 samples were tested for Nosema, with 98 (49%) positive for a Nosema infection. Of those, 37 (38%) contained *N.apis only infections*. 27 (13%) contained N ceranae only and the remaining 34 (35%) had both species).

Up to September 2014, 117 samples of bees had been examined for Nosema infections. 24 (21%) were positive for *Nosema*. 16 (14%) contained *N.apis*, 5 (4%) contained *N. ceranae* only and 3 (2.5%) were positive for both *Nosema* species.

Note, however, this should not be used as an indicator of prevalence, as samples were not spatially representative of colony distribution in Northern Ireland.

Bee Health Contingency Plan
The Bee Health Contingency Plan is reviewed annually and an updated version was published on the DARD Internet in November 2013.

Strategy for the Sustainability of the Honey Bee
The Strategy for the Sustainability of the Honey Bee was published in February 2011 and aims to achieve a sustainable and healthy population of honey bees for both pollination and honey production in the north of Ireland through strengthened partnership working between Government and Stakeholders. The Strategy confirms DARD's ongoing commitment to help protect and improve the health of honey bees and support the sector in its efforts to sustain and support beekeeping. The Ulster Beekeepers Association (UBKA) and the Institute of NI Beekeepers (INIB) have made a commitment to support the Strategy intentions. The Strategy is aimed at both policy makers and beekeepers, and importantly, identifies the roles and responsibilities of the different stakeholders in delivering its aims and outcomes. It seeks to address the current challenges facing beekeepers and provides a plan of action aimed at sustaining the health of honey bees and beekeeping in the north of Ireland for the next decade.

DARD

✉ ☎

The Strategy for the Sustainability of the Honey Bee can be viewed at:
http://www.dardni.gov.uk/strategy-for-the-sustainability-of-the-honey-bee.pdf

Beekeeping Courses at CAFRE
Preliminary Beekeeping Courses are organised and delivered by local Beekeeping Associations affiliated to the Ulster Beekeepers' Association (UBKA) at a range of venues around Northern Ireland. This Intermediate level Beekeeping course consists of three parts, Scientific, Practical, and Apiary Practical delivered over a period of at least two years and leading to the Federation of Irish Beekeepers' Associations (FIBKA) Intermediate Certificate of Proficiency in Beekeeping. Last year we trained 83 people over 7 courses in the Preliminary Beekeeping course and 43 people over 4 courses in the Intermediate Beekeeping course.

If you are interested in a beekeeping course please contact CAFRE Industry training at industry.trainingadmin@dardni.gov.uk or telephone 028 9442 6880

Thomas Williamson
Plant Health Inspection Branch, DARD
Sam Clawson
Bee Disease Diagnostics, AFBI
Andrew Adams
Farm Policy Branch, DARD
Kenny White
Short Course Manager, CAFRE

SG-AFRC
✉ ☎

The Scottish
Government

THE SCOTTISH GOVERNMENT AGRICULTURE, FOOD AND RURAL COMMUNITIES DIRECTORATE (AFRC) - RURAL PAYMENTS AND INSPECTIONS DIRECTORATE (RPID)

HEADQUARTERS
Lead Bee Inspector
Stephen Sunderland,
P Spur, Saughton House,
Broomhouse Drive,
Edinburgh, EH11 3XD
Tel: 0300 244 6672
e-mail: beesmailbox@
scotland.gsi.gov.uk

The Scottish Government (SG) is responsible for bee health Policy in Scotland. SG recognises the importance of a strong Bee health programme, not only for the production of honey, but also for the contribution that bees make to the pollination of many crop species and to the wider environment.

Honey bees are susceptible to a variety of threats, including pests and diseases, the likelihood and consequences of which have increased significantly over the last few years.

The Scottish Government takes very seriously any biosecurity threat to the sustainability of the apiculture sector and is working closely with colleagues in Food and Environment Research Agency's (Fera) National Bee Unit (NBU) to enable a more joined up approach to be taken throughout Great Britain on the issues surrounding bee health.

The Scottish Government has invested in the NBU's National web-based database for beekeepers "BeeBase" and actively encourages beekeepers to register onto the system. This service will provide bee health and disease outbreak information and will also assist Bee Inspectors in disease control. BeeBase also provides information on legislation, pests and disease recognition and control, interactive maps, current research areas and key contacts.

Beekeepers have a significant role to play in ensuring disease management and control within their own apiaries are in order as they have a legal obligation to report any suspicion of a notifiable disease or pest to the Bee Inspector at their local SGRPID Area Office. Bee Inspectors are responsible for the operation of The Bee Diseases and Pests Control (Scotland) Order 2007 in their area with duties including:-

194

- Inspection of apiaries for presence of statutory bee diseases
- Taking and delivering samples to SASA
- Issuing and removal of 'Standstill Notices'
- Issuing of 'Destruction Notices' and supervising destruction
- Informing beekeepers of treatment options for European Foul Brood (EFB), where appropriate
- Granting the option, after taking account of the recommendations of SASA, and carrying out treatment
- Carrying out follow-up inspections after destruction or treatment

SASA

- **Science and Advice for Scottish Agriculture (SASA)** is responsible for providing specialist technical support where duties include:
- Examination of submitted samples suspected of being infected with American Foul Brood, European Foul Brood, Small Hive Beetle (SHB) or *Tropilaelaps*.
- Reporting results on which pathogen or pest is present
- Recommending, in consultation with the Bee Inspector, the most suitable option, destruction or treatment, for each individual case of EFB.
- Where treatment is agreed, ordering supplies of the approved antibiotic
- Provision of a free diagnostic service to beekeepers to identify and confirm the presence of varroa.
- Maintaining technical liaison with NBU and providing technical documentation as required
- Providing training courses and demonstration material as required

SASA (SCIENCE AND ADVICE FOR SCOTTISH AGRICULTURE)
1 Roddinglaw Rd,
Edinburgh, EH12 9FJ

**BEE DISEASES,
FIONA HIGHET**
Plant Health Section
(0131) 244 8817

**PESTICIDE INCIDENTS,
ELIZABETH SHARP**
Chemistry Section
(0131) 244 8874

SG-AFRC
✉ ☎

PESTICIDE INCIDENTS

As part of the Wildlife Incident Investigation Scheme (WIIS), SASA undertakes analytical investigations into bee mortalities where pesticide poisoning may have been involved. Beekeepers should send samples of dead bees (200) direct to SASA, Chemistry Section, for analysis. In the case of major incidents, beekeepers are advised to contact their nearest SGRPID Area Office so that an early field investigation can be instigated.

THE FOLLOWING SCOTTISH GOVERNMENT RURAL PAYMENTS AND INSPECTIONS DIRECTORATE (SGRPID) STAFF ARE AUTHORISED BEE INSPECTORS. ALL BEE INSPECTORS HAVE EMAIL ADDRESSES AS "FIRSTNAME.SURNAME@SCOTLAND.GSI.GOV.UK"

EDINBURGH (HQ)
Steve Sunderland
(Lead Bee Inspector)
P Spur, Saughton House,
Broomhouse Drive,
Edinburgh, EH11 3XD
Tel: 0300 244 6672
Fax: 0300 244 9797

GRAMPIAN (INVERURIE AREA OFFICE)
Kirsteen Sutherland
Thainstone Court,
Inverurie, Grampian,
Aberdeenshire, AB51 5YA
Tel: (01467) 626247
Fax: (01467) 626217

SOUTHERN (DUMFRIES AREA OFFICE)
Angus Cameron
161 Brooms Road
Dumfries, DG1 3ES
Tel: (01387) 274400
Fax: (01387) 274440

CENTRAL (PERTH AREA OFFICE)
Steve Sunderland
Strathearn House
Broxden Business Park
Lamberkine Drive
Perth, PH1 1RZ
Tel: 01738 602043

HIGHLAND (INVERNESS AREA OFFICE)
Clem Cuthbert
Jane Thomson
Gordon Mackay
Longman House
28 Longman Road
Inverness, IV1 1SF
Tel: 01463 253 053

SOUTH EASTERN (GALASHIELS AREA OFFICE)
Angus MacAskill
Cotgreen Road
Tweedbank, Galashiels
Scottish Borders, TD1 3SG
Tel: (01896) 892400
Fax: (01896) 892424

SOUTH WESTERN (AYR AREA OFFICE)
John Smith
Russell House
King Street
Ayr
South Ayrshire
KA8 0BG
Tel: (01292) 291300
Fax: (01292) 291301

SCOTLAND'S RURAL COLLEGE (SRUC)

The Scottish Government supports a full-time apiculture specialist (Graeme Sharpe) who provides comprehensive advisory, training and education programmes for beekeepers throughout Scotland on all aspects of Integrated Pest Management, good husbandry (including the control of Varroa) and management practices. SAC also promotes the awareness of notifiable bee diseases and pests.

GRAEME SHARPE, APICULTURE SPECIALIST,
SAC Consulting, Scotland's Rural College (SRUC), Veterinary Services
John Niven Building
Auchincruive Estate
Ayr, Ayrshire
KA6 5HW
Tel:01292 525375

WWW.SCOTLAND.GOV.UK/TOPICS/APICULTURE/GRANTS/INSPECTIONS/BEEINSPECTIONS

BEEKEEPING METRIC CONVERTION TABLES

°CENT	FAHR		INCH	MM		INCH	MM		INCH	MM
0	32		$^1/_{25}$	1		$1^5/_8$	42		10	254
5	40		$^1/_{12}$	2		$1^{11}/_{16}$	43		$10^1/_4$	260
7	44		$^1/_8$	3		$1^9/_{20}$	48		$11^1/_4$	286
30	86		$^1/_{16}$	5		2	51		$11^1/_2$	292
34	92		$^1/_4$	6		3	76		$11^3/_4$	298
38	100		$^5/_{16}$	8		$4^1/_4$	108		12	305
43	110		$^3/_8$	9		$4^1/_2$	114		14	356
49	120		$^1/_2$	12.5		$4^3/_4$	121		$16^1/_4$	413
54	130		$^5/_8$	16		$5^1/_2$	140		$16^1/_2$	49
60	140		$^3/_4$	18		$5^3/_4$	146		17	431
62	144		$^7/_8$	22		6	152		$17^5/_8$	448
82	180		1	25		$6^1/_4$	159		$18^1/_8$	460
90	194		$1^1/_{16}$	27		$8^1/_4$	216		$18^1/_4$	483
100	212		$1^3/_8$	35		$8^3/_4$	223		20	508
			$1^9/_{20}$	37		$9^1/_8$	232		$21^1/_2$	546
			$1^1/_2$	38		$9^3/_8$	239		$21^3/_4$	552
						$9^9/_{16}$	246		22	559

INTERNATIONAL QUEEN MARKING COLOURS

YEAR ENDING	COLOUR	REMEMBER
1 & 6	WHITE	Will
2 & 7	YELLOW	You
3 & 8	RED	Raise
4 & 9	GREEN	Good
5 & 0	BLUE	Bees?

197

USEFUL TABLES

BOTTOM BEE-SPACE HIVES

No, of cells in brood box ─────────────────────────
Lug length (MM) ───────────────────────────
Frame spacing (mm) ─────────────────────
Frame size (mm) ─────────────────
No. frames ─────────────
Hive type ─────────

Hive type		No. frames	Frame size (mm)	Frame spacing (mm)	Lug length (MM)	No. of cells in brood box
National	BROOD	11	356 x 216	37	38	58000
	SUPER	10	356 x 140	42	38	36000
Modified Commercial	BROOD	11	406 x 254	37	16	75000
	SUPER	10	406 x 152	42	16	

TOP BEE-SPACE HIVES

No, of cells in brood box ─────────────────────────
Lug length (MM) ───────────────────────────
Frame spacing (mm) ─────────────────────
Frame size (mm) ─────────────────
No. frames ─────────────
Hive type ─────────

Hive type		No. frames	Frame size (mm)	Frame spacing (mm)	Lug length (MM)	No. of cells in brood box
Smith	BROOD	11	356 x 216	37	18	58000
	SUPER	10	356 x 140	42	18	36000
Langstroth	BROOD	10	448 x 232	35	16	68000
	SUPER	10	448 x 140	35	16	
Jumbo	BROOD	10	448 x 286	35	16	85000
	SUPER	10	448 x 140	35	16	
Modified Dadant	BROOD	11	448 x 286	37	16	93000
	SUPER	10	448 x 159	42	16	

USEFUL TABLES

CONVERSION FACTORS

TEMPERATURE

Fahrenheit > Celcius (Centigrade)	- 32, x 0.5555 ($^5/_9$)
Celcius > Fahrenheit	x 1.8 ($^9/_5$), + 32

WEIGHT

Ounces > Pounds	x 28.3495
Pounds > Grams	x 453.59237
Hundredweights > Kilograms	x 50.8
Grams > Ounces	'/. 28.3495
Kilograms > Pounds	x 2.2142

LENGTH

Inches > Centimetres	x 2.54
Yards > Metres	x 0.9144
Miles > Kilometres	x 1.609
Centimetres > Inches	x 0.3937
Metres > Yards	x 1.0936
Kilometres > Miles	'/. 1.609

AREA

Acres > Hectares	x 0.404686
Hectares > Acres	x 2.47105

VOLUME

Pints > Litres	x 0.5683
Gallons > Litres	x 4.546
Litres > Pints	x 1.7598
Litres > Gallons	x 0.21997

199

WORD SEARCH ANSWERS

BRASSICA
SALIX
TILIA
TRIFOLIUM
CALLUNA
ERICA
MALUS
PYRUS
PRUNUS
ONOBRYCHIS
SINAPSIS
CRATAEGUS
ACER
RUBUS
EPILOBOLIUM
VICA
FAGOPYRUM
TARAXACUM
ROBINIA
ALNUS
IMPATIENS
MELISSA
BORAGO
BUXUS
CYTISUS

CASTANIA
CROCUS
GALANTHUS
LIMONIUM
OENOTHERA
RIBES
CORYLUS
ALTHAEA
HEDERA
LAVANDULA
MEDICAGO
MAHONIA
ORIGANUM
RESEDA
MENTHA
PHACELIA
PAPAVER
LIGUSTRUM
SENECIO
SALVIA
SYMOPHORICARPOS
ARBUTUS
HELIANTHUS
THYMUS
ECHIUM

www.ingramcontent.com/pod-product-compliance
Lightning Source LLC
Chambersburg PA
CBHW072137270326
41931CB00010B/1786